Butterflies
of Surrey

Butterflies
of Surrey

GRAHAM A.COLLINS

SURREY WILDLIFE TRUST

Cover illustration: Silver-spotted Skipper, by David Dunbar

ISBN 0 9526065 0 X

British Library Cataloguing-in-Publication Data.
A catalogue record for this book is available
from the British Library.

© Graham Collins 1995
Surrey Invertebrate Atlas Project

First published 1995
by Surrey Wildlife Trust
School Lane, Pirbright, Woking, Surrey GU24 0JN.

FOREWORD

Surrey is one of the best counties for butterflies in Britain, but until now there has been no attempt to publish details of the distribution, ecology, and phenology of this group for the county. The only work hitherto produced covering the whole of the area is the *Victoria County History*, the butterfly section of which was published in 1902 and consists of little more than a list of species together with a handful of localities. This current work arose from a need to update the macrolepidoptera list for north-east Surrey (Evans, 1973), and the opportunity was taken to extend the coverage to the whole of the vice-county, at the same time providing for distribution maps as well as ecological data. The principal criterion for the list was scientific accuracy and consequently recorders of known competence were approached rather than accepting records from all and sundry. The coverage, in terms of dots on maps, is thus perhaps rather less comprehensive than some recent county lists, but the data is more scientifically valid.

This volume is the first in a series of county lists covering various insect orders or groups, in which the authors have given freely of their time and often made considerable financial input into the researching and recording of their subject. The costs of actually publishing their work are however far too great for any individual to bear and the successful production of this volume has only been possible through the great kindness of a number of bodies and individuals who are listed below.

Surrey County Council
British Entomological & Natural History Society
Butterfly Conservation (Surrey branch)
Croydon Natural History & Scientific Society
Corporation of London
Dominic Couzens

I am also extremely grateful for the assistance of the following: Martin Newman and staff (Surrey Wildlife Trust) for help in raising finances, supporting the Surrey Invertebrate Atlas Project and acting as publishers; Clare Windsor (Surrey Wildlife Trust) for the design and typesetting, and for organising the printing; Gail and Stephen Jeffcoate (Butterfly Conservation) for providing access to their database of records, arranging provision of slides from their membership, and helpful comments on individual species; Colin Plant (London Natural History Society) for providing records relating to the London area of Surrey; Dr Martin Warren for refereeing the species accounts; Roger Hawkins for proof-reading; and all the many individuals who have supported the project by providing records of butterflies – their names are listed separately.

The following have provided slides for the colour illustrations – Graham A. Collins, David Dunbar, Robert Edmondson, Barry Hilling, Tony Hoare, Gail Jeffcoate, Jim Porter, Paul Underwood, Mike Weller, Shirley White, Ken Willmott.

The distribution maps in this book have been produced by the program DMap (in its Windows version) written by Dr Alan Morton of Imperial College at Silwood Park. For further information about the program and its implementation in this case see Morton and Collins (1992).

GRAHAM A. COLLINS

CONTENTS

SURREY – THE STUDY AREA

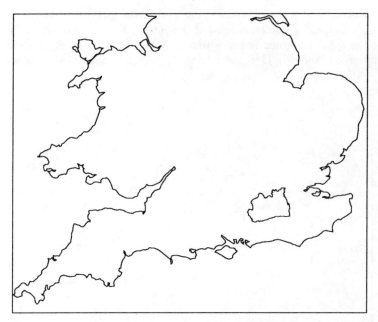

Surrey in relation to southern England

The boundary of the current county is a political boundary as ephemeral as the careers of the politicians responsible for running it, and consequently wholly unsuited to the recording of any biological group over any period of time. For this a stable boundary is necessary and exists in the form of the vice-county, a division originally proposed by H.C. Watson in 1852 in order to provide a set of unit areas of roughly similar dimensions for botanical recording. The system was rapidly adopted by botanists and supported by the forerunners of the Botanical Society of the British Isles and has since been in use by many zoologists. Since the introduction of this system the political boundary of Surrey has changed several times and is currently in danger of further alteration. The sense of using the fixed boundary of the vice-county can immediately be seen. The whole vice-county system is explained in text and maps in Dandy (1969).

The vice-county of Surrey differs from the current administrative county principally in its northern boundary which is marked by the course of the River Thames thus including the boroughs of the south-western quadrant of Greater London. The southern boundary differs slightly in that it runs almost east-west in the vicinity of Horley and so includes the area occupied by Gatwick airport which is currently in West Sussex, and on the western boundary an area of approximately one square kilometre to the south of the village of Batt's Corner and part of present day Surrey is excluded. The other major exclusion is the district of Spelthorne, which only became attached to Surrey in 1965, and in fact belongs principally to the vice-county of Middlesex.

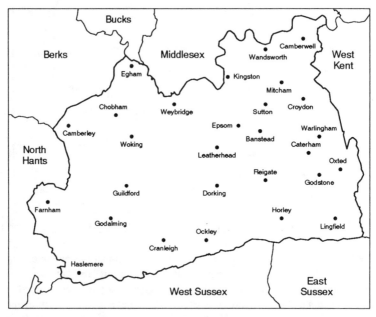

Surrey in relation to bordering vice-counties

GEOLOGY AND FACTORS DETERMINING DISTRIBUTION

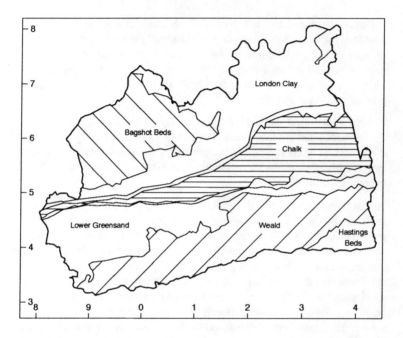

The geology of Surrey is composed of relatively young sedimentary rocks, laid down whilst our region was covered by water. Over tens of millions of years various strata built up, were raised into a dome, and by processes of erosion and weathering these strata were exposed at the surface. This has resulted in bands of these rocks running roughly east-west, with the older formations in the south of the county and the newer ones closer to the Thames. The strata have eroded at different rates, producing the hills of the North Downs and those around Leith Hill, which is the highest point in south-east England. The inclination of the strata, dipping to the north, gives these hills a shallow dip slope to the north and a steep scarp to the south.

The base geology determines the soil type above it, and, until recently, the type of land use. The sandy soils of the Lower Greensand, the Bagshot Beds and the Hastings Beds are acidic and very well drained, lacking many nutrients; when cleared and used for agriculture they were rapidly impoverished and became useless for anything but limited grazing, hence allowing the development of the heathland typical of western Surrey. The Chalk outcrop is basic, and particularly on the steep scarp slope unsuited to little but sheep grazing which has resulted in the short turf of downland. The clays of the Weald and the London Clay are heavy and poorly drained and again unsuited to early forms of agriculture; instead the original woodlands were utilized for timber and charcoal production and so formed the type of habitat favoured by such species as the Fritillaries. The base geology is modified by surface deposits, particularly in the flood plains of the river systems and on top of the chalk, allowing apparently anomalous habitats such as at Headley Heath.

Modern agricultural methods, encouraged by mechanical implements and Government grants, in addition to such changes in woodland management as the decline of coppicing and the growth of coniferous plantations, have considerably changed the Surrey landscape and the accelerating rate of change has taken its toll on our butterflies.

The distribution of butterflies within the county is limited, as much as anything, by the availability of their foodplants, and in the more specialist species by microclimatic features such as the local temperature. Many species are generalists, being found throughout the county, either using foodplants that are similarly undemanding in their requirements, or, as in the case of the Green Hairstreak, by using different foodplants in different areas. Other species such as the Grayling require relatively warm and dry habitats and so are limited to heathland and, formerly, the chalk. The most restricted species are the Fritillaries, certain Blues, and the Silver-spotted Skipper. The former group specialized in coppiced woodland or open grassland, where the ground was sufficiently open for the post-hibernation larvae to bask; of these, two species have become extinct in Surrey and a further two have declined drastically mainly as a result of habitat change. Several of the more local Blue species are limited by a choice of foodplant growing only on basic soils, or by their relationships with ant species, in particular the chalk form of the Silver-studded Blue which was probably entirely dependent on an ant species which itself required short turf, and was very likely lost as an aftermath of myxomatosis.

HISTORY OF PUBLISHED WORKS

The only list of lepidoptera to include the whole of the county is that which appears in the *Victoria County History* of 1902 and is indeed little more than a list of species together with a handful of sites from which they were then known. The butterfly section however makes interesting reading as the author considered the county to be considerably poorer than many of the neighbouring ones and many species to be significantly rarer than is the case today. This contrasts rather with the "things were much better in the old days" syndrome which affects many modern commentators, and shows that in the matter of increase and decline of species one must take the long term view. Subsequent recording, in increasing detail and in improved presentation, has been restricted to more circumscribed areas within the vice-county.

The London area, defined as an area of 20 miles radius with its centre at St. Paul's Cathedral and utilized by the London Natural History Society, was the subject of various papers by Baron de Worms which appeared in the *London Naturalist* from 1949 to 1959. This area and the vice-county of Surrey overlap, the arc of its boundary passing through Weybridge, Mickleham, Nutfield and Limpsfield; records therein are subdivided into the relevant vice-counties.

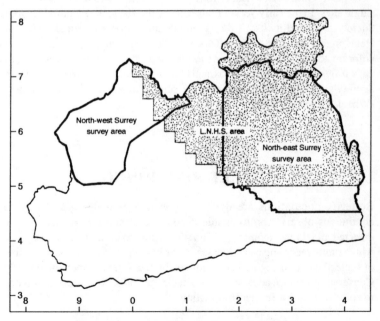

Survey areas used in previous publications

Around the same time Russell Bretherton was compiling his list of the macrolepidoptera of north-west Surrey, an area he defined as being bounded to the west and north by the county boundaries and to the south and east by the railways from Guildford to, respectively, Ash Vale and Woking through to Weybridge.

The next major list to appear was that for Croydon and north-east Surrey (Evans, 1973), an area broadly similar in size to the north-west Surrey survey. This list was described at the time by its reviewer as "the best authenticated and most attractively presented local list we have seen", and included information such as voltinism, habits and for many species a complete listing of records. In addition a limited number of distribution maps were provided. It was this work and the guidance of its author, Ken Evans, which steered me through my formative years in entomology, and the desire to update and expand the list led directly to the formation of the current recording scheme.

More recently the London list of de Worms has been updated by Colin Plant, to include many more modern records together with tetrad distribution maps for all the species.

In addition to these lists which cover a major portion of the county, there have been a number of lists covering much more restricted areas or even single sites. In the early years of this century the Haslemere Natural History Society produced its list of the Lepidoptera occurring within six miles of Haslemere (Oldaker, 1913). This is little more than a list, and, most unfortunately, gives no indication to which of the three possible vice-counties the various records pertain. Bookham Common, the London Natural History Society's special survey area, was the subject of a list published in 1955 (Wheeler, 1955).

FUTURE RECORDING

Butterfly populations are dynamic, constantly changing and require continuous observation to enable declines and increases in range to be detected and threats monitored. This survey should act as a baseline against which future recording can be compared. Following its publication, recording of butterflies in Surrey will be continued by the Surrey branch of Butterfly Conservation whose address is given in the appendix. Recorders are invited to contact them for further information.

PUBLISHED SURREY LISTS

Goss, H., 1902.
> Butterflies and Moths. *A History of the County of Surrey,* **3** *Zoology. (VICTORIA COUNTY HISTORY.)* Constable, London.

de Worms, C.G.M., 1950.
> The Butterflies of London and its Surroundings. *Lond. Nat.* **29**:46-80.

Bretherton, R.F., 1957.
> A List of the Macrolepidoptera and Pyralidina of north-west Surrey. *Proc. S. Lond. ent. nat. Hist. Soc.* **1955**:94-151.

Bretherton, R.F., 1965.
> Additions to the List of Macrolepidoptera and Pyralidina of north-west Surrey. *Proc. S. Lond. ent. nat. Hist. Soc.* **1965**:18-30.

Evans, L.K., and Evans, K.G.W., 1973.
> A Survey of the Macrolepidoptera of Croydon and north-east Surrey. *Proc. Croydon Nat. Hist. Sci. Soc.* **XIV**:273-408.

Plant, C.W., 1987.
> *The Butterflies of the London Area.* London Natural History Society.

CURRENT PUBLISHED LISTS FOR ADJOINING VICE-COUNTIES

KENT

Chalmers-Hunt, J.M., 1960-61.
>*The Butterflies and Moths of Kent*, **1**. Arbroath and London.

Philp, E.G., 1993.
>*The Butterflies of Kent*. Kent Field Club, Sittingbourne.

SUSSEX

Pratt, C., 1981.
>*A History of the Butterflies and Moths of Sussex*.
>Booth Museum, Brighton.

NORTH HANTS

Goater, B., 1974.
>*The Butterflies and Moths of Hampshire and the Isle of Wight*. Classey, Faringdon.

Goater, B., 1992.
>*The Butterflies and Moths of Hampshire and the Isle of Wight: additions and corrections*. Joint Nature Conservation Committee.

BERKS

Baker, B.R., 1994.
>*The Butterflies and Moths of Berkshire*. Hedera Press, Uffington.

BUCKS

Ansorge, E., 1969.
>*The Macrolepidoptera of Buckinghamshire*. Bucks Archaeological Society, Aylesbury.

BERKS/BUCKS

Asher, J., 1994.
>*The Butterflies of Berkshire, Buckinghamshire and Oxfordshire*. Pisces.

MIDDLESEX

Plant, C.W., 1987.
>*The Butterflies of the London Area*. London Natural History Society. [Includes the whole of VC21 Middlesex, plus part of Surrey].

COLLECTING AND THE LAW

Although many conservationists are now becoming more enlightened, the topic of insect collecting still arouses considerable passions. There are two arguments against collecting; the moral one, and whether any harm is done to the population as a whole by removing individual specimens. Morals are a personal matter and one individual has no right to inflict them upon another, so the only argument that needs consideration is the potential of damage to butterfly populations. The population dynamics of insect species are such that a large number of eggs are laid by one female so that the often considerable loss of the early stages by predation and disease still results in the replacement of the previous generation. Slightly simplified, if we consider that one male together with the female is responsible for the next generation then it is necessary for only two individuals to survive if the population is to remain stable. For many species this is only a few percent of the number of eggs laid. In addition one male may fertilize several females. Thus the removal from the wild of small numbers of early stages or even adult males has an insignificant affect on the population as a whole, and as long as the population is sufficiently large the loss of a few females also has no effect.

As long as the JCCBI's code for collecting is followed there are no species in Surrey that will be endangered by amateur collectors. Commercial breeders, who remove much larger numbers than the amateur ever would, can however be a problem and this is catered for by the Wildlife and Countryside Act (1981) which lists some two dozen species which cannot be traded without a licence. Recent prosecution of a dealer shows that the Act can and will be used. There are no butterfly species on schedule five, which affords them complete protection, that occur in Surrey.

Many insects, including a few butterfly species, need examining at close quarters for accurate identification, and many of them need to be killed and examined microscopically to make this identification, including even some of the larger moths. The numbers that it is necessary to kill are but a minute proportion of the population and our gain in knowledge far outweighs the loss of a few specimens. Recording of such species as the two small Skippers, and the Small and Green-veined Whites, is most easily accomplished by netting them; this allows close examination, after which they can be released unharmed.

The other aspect of the Wildlife and Countryside Act which concerns us is the prohibition of the release into the wild of insects of a kind not occurring naturally here. The indiscriminate release of specimens has become such a problem that there are proposals to include the species which cannot be traded on a list of species which it would be illegal to release.

THE CHANGING STATUS OF SURREY'S BUTTERFLIES

Butterfly populations are rarely static and fluctuate over both the short and long-term, as a result of habitat change, climate, and host-parasite interactions. When comparing with the time of, say, the Victoria County History (published in 1902), it is clear that considerable amounts of land have been lost to development, especially in the London suburbs, and it is also the case that habitats have changed as a result of changing woodland management and particularly "improved" farming methods. Some of these changes have been gradual and some more sudden.

Other factors affecting butterfly populations are the advent of Dutch-elm disease and the severe changes to the rabbit population brought about by myxomatosis. The only species dependent on elm, the White-letter Hairstreak, has survived the loss of elms very well, still being a widespread and frequently common species. The effect on downland of the loss of rabbits has been more serious, and only in recent years, with rabbit numbers increasing again and the re-introduction of sheep grazing to the downs, has the habitat been restored to any appreciable degree. Downland butterflies such as the Chalk-hill Blue, the Adonis Blue and the Silver-spotted Skipper undoubtedly declined during this period, although the Skipper at least has apparently regained lost ground. Other butterflies such as the Silver-studded Blue and the Grayling have become extinct on the chalk, although both are still fairly common on heathland. A further species, the Marbled White, was probably extinct in Surrey before myxomatosis, but subsequent introductions have fared better than the original residents, and this may be due to the increase in lush grass in its chalky habitats.

Woodland butterflies have also altered their statuses over the years, some favourably as in the case of the White Admiral and, to a certain extent, the Wood White, and some disastrously such as the two species of Pearl-bordered Fritillary which have declined seriously, and the High Brown Fritillary which has been extinct for some time.

The table opposite attempts to chart the change in status of the Surrey butterflies since the turn of the century.

1. Species which have become extinct since 1900

Plebejus argus cretaceus	Silver-studded Blue	chalk subspecies only
Argynnis adippe	High Brown Fritillary	
Eurodryas aurinia	Marsh Fritillary	
(Nymphalis polychloros	Large Tortoiseshell	possibly only temporary resident)

2. Species whose decline is a matter for concern

Lysandra bellargus	Adonis Blue
Boloria selene	Small Pearl-bordered Fritillary
Boloria euphrosyne	Pearl-bordered Fritillary
Lasiommata megera	Wall

3. Species which have declined somewhat

Hesperia comma	Silver-spotted Skipper	currently increasing
Erynnis tages	Dingy Skipper	
Pyrgus malvae	Grizzled Skipper	
Lysandra coridon	Chalk-hill Blue	decline in numbers
Hamearis lucina	Duke of Burgundy	
Argynnis aglaja	Dark Green Fritillary	currently very uncommon
Hipparchia semele	Grayling	lost from the chalk

4. Species which have increased considerably

Thymelicus lineola	Essex Skipper	since 1950s
Leptidea sinapis	Wood White	increase and subsequent decline
Ladoga camilla	White Admiral	
Apatura iris	Purple Emperor	
Polygonia c-album	Comma	
Pararge aegeria	Speckled Wood	
(Melanargia galathea	Marbled White	as a result of introduction)

INTRODUCTIONS

The artificial introduction of butterfly populations has a long history but in recent years has become both more popular and more scientifically applied, if not less controversial. Several recent publications set out to broadly justify such introductions (i.e. Emmet, 1989, and Oates and Warren, 1990), although it should be understood that these authors are in favour of the practice. In my view introductions should only be attempted in extreme cases, at sites where the butterfly has previously occurred, and where there is no chance of natural recolonization. In Surrey the only butterflies which even merit vague consideration for such procedures are the two species of Pearl-bordered Fritillaries. Other species such as the Wall, which admittedly has declined drastically in recent years but has the full potential to recolonize the county naturally, should not be interfered with. The most important consideration is that any releases of butterflies are fully documented, and that they are made with the approval of competent conservationists as indiscriminate releases are just as likely to damage wild populations as to enhance them.

Species which are extant in the county and have been introduced are considered in the species accounts later in the book; other species for which documented introductions have taken place but no longer occur in the county are listed below.

Mellicta athalia (**Heath Fritillary**). Oates and Warren (1990) document several releases of this species in Surrey. In 1926 it was introduced into an unspecified wood and examples were seen the following year. In 1958 it was decided by the then Lepidoptera Protection Committee to introduce it into woods at Oxshott. This involved not only the introduction of the butterfly but also the prior introduction of the foodplant which did not occur naturally at the site. The outcome was unknown. More recently a few releases have been attempted in the woods around Chiddingfold, where the habitat and management are totally unsuitable. The Heath Fritillary is afforded schedule five status under the Wildlife and Countryside Act (1981) and as it cannot be collected in any of its stages any unofficial attempts at introduction would also be illegal.

Papilio bianor. Large numbers of this Asian Swallowtail were released at Witley and Bagshot, and at other areas in south-east England, in June 1917. In Hampshire at least it was seen again in 1918 having presumably survived the winter, but not surprisingly died out soon after.

LIST OF RECORDERS

The following people have supplied records to the scheme and their data has been used to compile the accounts of the species together with the distribution maps. In order to save space when individual records are listed the recorders initials have frequently been used and are listed below.

M.Adler
P.J.Baker (PJB)
D.W.Baldock (DWB)
A.J.Baldwin (AJB)
R.Barnett
J.Bebbington
M.Bennett
P.Beuk
G.Blaker
R.F.Bretherton (RFB)
N.Brown
P.Cattermole (PC)
B.Chesney (BC)
S.H.Church (SHC)
P.Churchill
J.Clarke (JC)
D.Coleman (DC)
G.A.Collins (GAC)
G.B.Collins (GBC)
P.F.Collins (PFC)
P.A.Cordell
J.Cranham
R.A.Cramp (RAC)
J.V.Dacie (JVD)
T.J.Daley (TJD)
D.Dell
C.G.M.de Worms (CGMdeW)
D.Dunkin
M.Ellis (ME)
E.Emmett (EE)
M.Enfield
K.Everard
P.Farrant
R.Fairclough (RF)
P.Follett (PF)
S.W.Gale (SWG)
M.Gascoigne-Pees (MG-P)
G.Geen (GG)

B.Gerrard
P.Grove
A.& O.Hall (A&OH)
A.J.Halstead (AJH)
C.Hart (CH)
M.S.Harvey (MSH)
R.Hastings
J.Hatto
R.D.Hawkins (RDH)
B.Hilling
A.Hoare
P.J.Holdaway (PJH)
J.D.Holloway (JDH)
S.F.Imber (SFI)
G.Jeffcoate (GJ)
A.Jones
A.M.Jones (AMJ)
R.A.Jones
Juniper Hall Field Centre (JHFC)
A.Kerr
S.Kett
D.C.Lees (DCL)
A.Lickley
I.D.MacFadyen (IDMacF)
H.Mackworth-Praed (HM-P)
P.A.Martin (PAM)
R.F.McCormick (RFMcC)
I.Menzies
J.L.Messenger (JLM)
E.Moore (EM)
R.K.A.Morris (RKAM)
M.Parsons (MP)
S.Paston
A.Petrie
S.C.Pittis (SCP)

C.W.Plant (CWP)
J.Pontin
J.Porter (JP)
S.Price
K.Reel
A.Reid
M.Reid
G.Revill
J.Ruffin
J.T.Scanes (JTS)
P.J.Sellar (PJS)
I.G.Shearer (IGS)
A.Shelley
R.G.Shotter (RGS)
D.Smith
R.Smith
T.Smithers
J.B.Steer (JBS)
P.M.Stirling (PMS)
I.Stone
Surrey Biological Records Centre (SBRC)
G.Taylor
D.A.Trembath (DAT)
P.Underwood
R.Wason
A.S.Wheeler (ASW)
H.Whiting
B.Whitty
E.H.Wild (EHW)
A.A.Wilson (AAW)
A.Wingrove
T.G.Winter (TGW)
E.Wood
M.Woolven
E.Young

EXPLANATION OF SPECIES ACCOUNTS

The following accounts contain details of all the species of butterfly recorded in Surrey as reported to me or published in the various journals or textbooks. Each species has a summary of its status, habitat preferences, distribution within the county and frequency, its voltinism and flight periods, and the feral foodplants from which its larvae have been noted. The more localized species also have appended a list of all the individual records received, arranged more or less geographically. Rarer migrants have all records arranged chronologically. Unless otherwise stated all details refer to Surrey data.

The following definitions are used:

STATUS

Resident Species which have been noted breeding or have occurred in sufficient numbers to indicate breeding within the last 20 years, and whose existence does not rely on migration. This is the only group for which conservation measures need be considered.

Extinct Species formerly resident, but which have not been recorded within the last 20 years. While there are a few British species which have been lost for a hundred years before being rediscovered, a period of 20 years absence in a well worked county is considered sufficient grounds for considering a species extinct.

Migrant Species which occur in the county solely by virtue of their passage from continental Europe or North Africa, such passage being, as far as can reasonably be determined, under their own power. These species may breed, some regularly, but cannot survive the winter except on rare occasion and would die out without continual replenishment from foreign stock.

Vagrant Species whose presence in the county cannot be explained by either residency or migration. This includes species that may have been transported with garden plants, agricultural produce, or even unwittingly by other entomologists, as well as species not generally considered to be migratory. Tropical species, clearly arriving here by the agent of man are not considered within the scope of a local list.

Introduction   Species deliberately introduced by man. They may have no other claim to be Surrey species, or they may supplement native populations; in cases where these introductions are insufficiently documented it is often not possible to be sure.

Status uncertain   A few species are insufficiently recorded to be considered resident, but on balance may be.

DISTRIBUTION

Widespread   Species occurring throughout most of the county, without particular preference for habitat or soil type.

Restricted   Species occurring fairly widely but nevertheless absent from large areas often as a result of habitat requirement.

Local   Species recorded from only a few sites, although these sites themselves may be widespread.

Very local   Species recorded from very few sites, often in the same general area or geological formation.

FREQUENCY

The frequency of a species is necessarily subjective, depending on a particular recorder's methods and experience. Some species are rarely observed as adults but can be found much more easily and in greater numbers as larvae. The frequency is thus estimated from all sources and is given on a sliding scale as follows:

Common / Fairly common / Uncommon / Scarce / Rare

VOLTINISM

The voltinism or "broodedness" of many species is complicated and not easy to determine due to the comparative paucity of larval records. It may be **univoltine**, with a single brood of adults occurring in any one year, or **bivoltine**, with two broods, offspring of the first contributing wholly and completely to adults of the second. In some species the second brood is only partial, some pupae lying over to produce adults of the first brood of the following year; the offspring of the second brood may survive to the following year or they may die from lack of food and cold as the winter progresses. Species in which the second brood is only very small and occasional are considered to be univoltine, but these exceptions are mentioned in the text.

FLIGHT PERIOD

The flight period stated is that time in which most of the adults are to be found and refers to Surrey data obtained during the last 15 years. Extremely early or late individuals are excluded.

FOODPLANT

The list of foodplants refers wholly to Surrey data, and mostly includes plants that are native, agricultural crops, or escaped garden plants. Plants of domestic gardens are generally excluded except in cases where a garden plant forms a major, or indeed the only, recorded foodplant. The names of plants are given in English; a list of scientific names is given in the appendices. The following definitions are used:

Foodplant   where listed indicates that it has been utilized as a foodplant during the last 15 years.

[Foodplant]   a foodplant in brackets indicates that there is insufficient evidence to indicate that the food was used, but where probability suggests that it has – for example where a pupa has been dug from the base of a particular tree.

(Foodplant)   a foodplant in parentheses indicates that there are old records, but that the particular plant has not been recorded as being used during the last 15 years.

REFERENCES TO PUBLICATIONS

Some of the information in the book has been derived from data already published in other books and the entomological journals. Where the reference contains more detailed or related information, the full source is listed in the reference section of the appendix. Where, however, the reference is merely a record published in one of the journals rather than one which has been submitted to me directly, an abbreviated form of the reference is given in the text. The abbreviations and the full journal names are as follows:

Ent. Rec. *Entomologist's Record and Journal of Variation.*

Ent. *The Entomologist.*

Ent. Gaz. *Entomologist's Gazette.*

E.M.M. *Entomologist's Monthly Magazine.*

L.N. *London Naturalist.*

Proc. BENHS *Proceedings and Transactions of the British Entomological and Natural History Society.*

Proc. SLENHS *Proceedings and Transactions of the South London Entomological and Natural History Society.*

E.W.I. *Entomologist's Weekly Intelligencer.*

MBGBI *Moths and Butterflies of Great Britain and Ireland* (i.e. Emmet, 1989).

VCH *Victoria County History* (i.e. Goss, 1902).

DISTRIBUTION MAPS

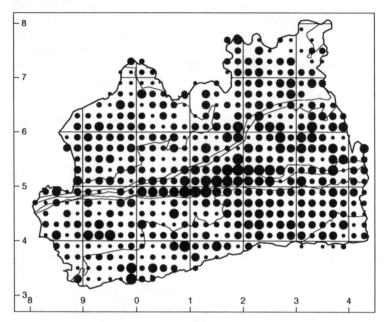

Coverage of butterfly records 1980-94. The larger the dot, the greater the number of species recorded in that tetrad.

The distribution maps accompanying the resident species show from which tetrads the butterflies have been recorded since 1980, a tetrad being an area 2 km square. The national grid lines are also shown, allowing the location of sites using the grid references given in the gazetteer, as are the boundaries of the main geological formations. (See map on page 3.)

A round symbol represents an adult butterfly, and a square one a record of one of the early stages. It has not proved possible to visit every tetrad in the county, and the absence of a dot from any particular square should not be taken to imply that a particular species does not occur there; rather the maps should be viewed as a graphical representation of the distribution as described in the species accounts and lists of records. The map above shows recording coverage, and illustrates a certain amount of recorder bias as well as highlighting the better areas for butterflies. This bias should be borne in mind when consulting the individual distribution maps. Maps for migrant species are not included.

The maps were prepared from information held in a species database using the program DMap (for Windows) written by Alan Morton of Imperial College.

CHECKLIST OF SURREY BUTTERFLIES

> 1. Not recorded during the survey period, 1980-94.
> 2. Probably reinforced by migration.

HESPERIIDAE – Skippers

Hesperiinae

Carterocephalus palaemon	Chequered Skipper	doubtful[1]
Thymelicus sylvestris	Small Skipper	resident
Thymelicus lineola	Essex Skipper	resident
Hesperia comma	Silver-spotted Skipper	resident
Ochlodes venata	Large Skipper	resident

Pyrginae

Erynnis tages	Dingy Skipper	resident
Carcharodus alceae	Mallow Skipper	vagrant[1]
Pyrgus malvae	Grizzled Skipper	resident

PAPILIONIDAE – Swallowtails

Papilioninae

Papilio machaon	Swallowtail	extinct resident and vagrant

PIERIDAE – Whites

Dismorphiinae

Leptidea sinapis	Wood White	resident

Coliadinae

Colias hyale	Pale Clouded Yellow	migrant[1]
Colias croceus	Clouded Yellow	migrant
Gonepteryx rhamni	Brimstone	resident

Pierinae

Aporia crataegi	Black-veined White	extinct resident
Pieris brassicae	Large White	resident[2]
Pieris rapae	Small White	resident[2]
Pieris napi	Green-veined White	resident
Pontia daplidice	Bath White	migrant[1]
Anthocharis cardamines	Orange-tip	resident

LYCAENIDAE – Hairstreaks, Coppers, and Blues

Theclinae

Callophrys rubi	Green Hairstreak	resident
Thecla betulae	Brown Hairstreak	resident
Quercusia quercus	Purple Hairstreak	resident
Satyrium w-album	White-letter Hairstreak	resident
Satyrium pruni	Black Hairstreak	introduction

Lycaeninae

Lycaena phlaeas	Small Copper	resident

Polyommatinae

Lampides boeticus	Long-tailed Blue	migrant/vagrant
Cupido minimus	Small Blue	resident
Plebejus argus argus	Silver-studded Blue	resident
Plebejus argus cretaceus	Silver-studded Blue	extinct resident
Aricia agestis	Brown Argus	resident
Polyommatus icarus	Common Blue	resident
Lysandra coridon	Chalk-hill Blue	resident
Lysandra bellargus	Adonis Blue	resident
Cyaniris semiargus	Mazarine Blue	extinct
Celastrina argiolus	Holly Blue	resident

Riodininae

Hamearis lucina	Duke of Burgundy	resident

NYMPHALIDAE – Tortoiseshells, Fritillaries, and Browns
(see main text)

Limenitinae

Ladoga camilla	White Admiral	resident

Apaturinae

Apatura iris	Purple Emperor	resident

Nymphalinae

Vanessa atalanta	Red Admiral	migrant
Cynthia cardui	Painted Lady	migrant
Aglais urticae	Small Tortoiseshell	resident[2]
Nymphalis polychloros	Large Tortoiseshell	status uncertain
Nymphalis antiopa	Camberwell Beauty	migrant
Inachis io	Peacock	resident

Nymphalinae (cont)

Polygonia c-album	Comma	resident
Araschnia levana	Map Butterfly	vagrant

Argynninae

Boloria selene	Small Pearl-bordered Fritillary	resident
Boloria euphrosyne	Pearl-bordered Fritillary	resident
Boloria dia	Weaver's Fritillary	introduction
Argynnis lathonia	Queen of Spain Fritillary	migrant[1]
Argynnis adippe	High Brown Fritillary	extinct resident
Argynnis aglaja	Dark Green Fritillary	resident
Argynnis paphia	Silver-washed Fritillary	resident

Melitaeinae

Eurodryas aurinia	Marsh Fritillary	extinct resident/ introduction
Melitaea cinxia	Glanville Fritillary	extinct resident

Satyrinae

Pararge aegeria	Speckled Wood	resident
Lasiommata megera	Wall	resident
Melanargia galathea	Marbled White	extinct resident/ introduction
Hipparchia semele	Grayling	resident
Hipparchia fagi	Woodland Grayling	vagrant[1]
Arethusana arethusa	False Grayling	vagrant[1]
Pyronia tithonus	Gatekeeper	resident
Maniola jurtina	Meadow Brown	resident
Aphantopus hyperantus	Ringlet	resident
Coenonympha pamphilus	Small Heath	resident

Danainae

Danaus plexippus	Monarch	migrant[1]

HESPERIIDAE – Skipper Butterflies

The Skippers belong to the superfamily Hesperioidea, which puts them on an equal taxonomic footing with all the other butterflies which belong to the Papilionoidea. Taxonomically they are in a group somewhat intermediate between most butterflies and the moths, in the British species the difference in particular being seen in the resting posture which in some species is distinctly moth-like. Many tropical species are also crepuscular. The males have patches of scent-scales (androconia) which in the Hesperiinae are arranged as an oblique black line across the forewing, serving immediately to distinguish the sexes. In the Pyrginae these androconia are located in a fold in the costa of the forewing where they are less obvious. The Silver-spotted Skipper overwinters as an ovum and the Grizzled Skipper as a pupa; all the remaining British species overwinter as larvae, in varying stages of development from the Essex Skipper as a fully developed larva remaining within the eggshell to the Dingy Skipper which hibernates fully grown within a chamber in which it later pupates. The larvae are tapered and have the first segment much narrower than the head giving the appearance of a neck (plate 8). They live amongst spun leaves of their foodplant, emerging to feed at night when they may be found by sweeping or searching, the Hesperiinae being restricted to grasses. Pupation takes place within a cocoon of spun leaves. There are six resident species in Surrey.

Thymelicus sylvestris (Poda, 1761)   PLATES 1,8   **Small Skipper**

Resident; grassland, heathland, open woodland; widespread and common.

Univoltine; July and August.

Foodplant – (timothy grass).

The Small Skipper is a widespread species which occurs commonly in open grassy areas across the county. Confusion with the next species is likely unless the undersides of the antennae are examined – they are jet-black in the Essex Skipper and orange in the Small Skipper; occasional examples having dusky

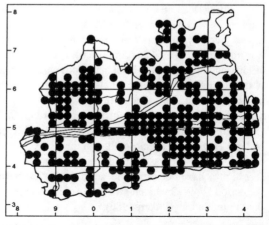

antennae are invariably *sylvestris*. Specimens at rest or feeding can often be inspected, but the quickest method is to catch them. Worn or badly damaged specimens can be confirmed by examination of the genitalia (see MBGBI 7(1):58). The two *Thymelicus* species frequently occur together but often in differing numbers suggesting a slight difference in habitat preference. This situation is complicated by the Small Skipper emerging slightly earlier than the Essex due to the larva hatching in the late summer and overwintering in this stage rather than as an egg. The Small Skipper seems to exhibit a preference for more established grassland, and is commoner than the following species in more shaded areas such as woodland rides. The larvae can be swept readily from grasses after dark in the spring and can be distinguished by the head capsule – green in the Small Skipper and brown with paler vertical stripes in the Essex.

Thymelicus lineola (Ochsenheimer, 1808)   PLATE 1   **Essex Skipper**

Resident; grassland, heathland, open woodland; widespread and common.

Univoltine; July and August.

Foodplant – grasses.

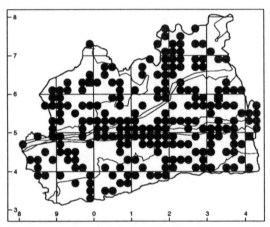

The Essex Skipper, like its congener the Small Skipper, occurs commonly in more open habitats throughout Surrey. As a British species it was not recognized until 1888 from specimens captured in Essex, and after publication of its discovery earlier examples were found in collections but all coming from the south-eastern coastal counties. An early London list (Buckell and Prout, 1898-1901) gave as localities: Croydon, Dulwich and Wimbledon Common. In north-west Surrey a specimen was taken near Bagshot on 3.7.1927 (Heslop, *Ent.* **61**:139) which was the only record given by Bretherton in his list and its supplement (Bretherton, 1957 and 1965). It was unrecorded in the Charterhouse list (Perrins, 1959), the London area list of de Worms (1950), and on the list for Bookham Common (Wheeler, 1955), although in the north-east of the county it had become regular in the Selsdon area by 1950. Evidence suggested that it had increased considerably in the early 1960s (Evans, 1973), although the picture is somewhat confused by lack of proper recording and the apparent assumption that any small skipper butterfly was bound to be *sylvestris*. Most areas of suitable habitat are inhabited by both *Thymelicus* species although the Essex Skipper seems to predominate in recently created habitat such as roadside verges suggesting a possible mechanism by which its spread may have occurred. Indeed in north-west Surrey it was virtually unrecorded until the period 1969-72 when it swarmed on the swathe of land cleared for the construction of the M3 motorway (Baker, 1986).

Hesperia comma (Linnaeus, 1758)   PLATE 1   Silver-spotted Skipper

Resident; chalk downland; local but fairly common.

Univoltine; August.

Foodplant – sheep's-fescue.

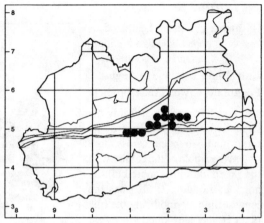

The Silver-spotted Skipper is restricted in Surrey to the scarp slope of the North Downs, principally between Hackhurst Downs in the west and Colley Hill in the east. Within this limited area the butterfly is not uncommon flying over short turf in early August. The butterfly has a requirement for high temperatures, being inactive on cold or overcast days. An essential habitat criterion is for patches of bare ground on which the adult will bask, and in particular around which the females will oviposit. Such patches are often formed by trampling by livestock and by material ejected from rabbit burrows; close grazing of the turf is also effected by these animals. Nationally the Silver-spotted Skipper is a rare species, being afforded RDB3 status, and Surrey is one of its strongholds having roughly half of the total British population. It was undoubtedly once more widespread in Surrey as, in addition to its current sites, Newman (1874) gives "on all the chalk downs in the Croydon district". Twenty years later Barrett (1893) says "at Croydon it has been exterminated", while Tutt (1905) adds Park Downs. There is, however, evidence of a recent modest expansion following recovery of the rabbit population. The usual foodplant would appear to be sheep's-fescue but the females have also been observed laying on wood false-brome and tor grass although it is very doubtful whether the larvae are able to complete their growth on these species (Willmott, *Ent. Rec.* **103**:247).

RECORDS – **Colekitchen Down**, 3-27.8.94 (GJ); **Hackhurst Downs**, 2.9.90 (RDH), 3.8.94(GJ); **White Downs**, 21.8.85 (AJH), 16.8.91 (GAC), 31.7.94 (B. Hilling); **Westcott Downs**, 9.8.84 (GAC), 13.8.85 (DWB), 21.7.94 (GJ); **Ranmore**, 13.8.93 (PFC); **Box Hill**, 12.8.81 (BC), 14.8.85 (PFC), 13.8.84, 19.8.87, 9.8.91, 7.8.93 (GAC), 23.7.94 (A. Kerr); **Headley Down** 4.8.94 (GJ); **Headley Warren**, 8.8.93 (HM-P); **Betchworth**, 8.8.83 (RFMcC), 7.8.93 (GAC); **Dawcombe**, 25.7.92 (RDH), 6.8.94 (GJ); **Buckland Hills**, 23.8.81 (RDH), 30.7.94 (GJ); **Juniper Hill**, 18.8.85 (RDH), 30.7.94 (GJ); **Colley Hill**, 23.8.84 (BC), 28.8.85, large colony (DWB), 1.8.94 (A. Reid).

Ochlodes venata (Bremer and Grey, 1852)   PLATE 1   **Large Skipper**
ssp. *faunus* (Turati, 1905)

Resident; grassland, heathland, open woodland; widespread and common.
Univoltine; late June to August.
Foodplant – [wood false-brome].

The Large Skipper is probably the most widespread member of its family and is frequent across more or less the whole of the county with the exception of the more built-up areas of London. It also has the longest flight period being the first of the Hesperiine Skippers to appear and still being on the wing well into

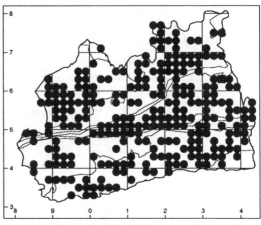

August, overlapping with the Silver-spotted Skipper with which it may be confused. Population numbers though are rather lower than with the Small and Essex Skippers, whose habitat it often shares; the greater size and different wing pattern of this species enable it to be easily separated from them. The larvae spend most of the time enclosed in a spun tube of grass and are only occasionally found by sweeping or searching, the feral foodplant not having been recorded although eggs have been found on wood false-brome on the chalk (GJ).

Erynnis tages (Linnaeus, 1758)   PLATE 2   **Dingy Skipper**
ssp. *tages* (Linnaeus, 1758)

Resident; chalk grassland, open woodland; restricted but fairly common.
Univoltine; late May and June.
Foodplant – bird's-foot trefoil.

In Surrey the Dingy Skipper is principally a species of the chalk and is widespread on this formation, especially on the scarp of the North Downs, and locally fairly common. Elsewhere it is less frequently recorded in areas of open woodland, mainly on the weald. A few records have also come from the Bagshot sand area: Ash, 4.6.83 (RDH), 1994 (D. Dell); Bisley Ranges, 7.6.93 (J. Pontin); and

Chobham Common, 7.6.88 (GAC); it is clearly no more than occasional in this area. There is evidence of a recent decline as de Worms (1950) lists it from Esher, and Wimbledon and Mitcham commons, all sites where it no longer occurs. Second brood examples are distinctly rare, being noted at Westcott Downs, 22.7.1989 (GAC), Dawcombe, 6.8.1994 (GJ) and another as late as 20.9.1975 at Brockham Warren (PJH). This species is particularly moth-like, especially when roosting, and hence rather easily overlooked; there is also a superficial resemblance to certain species of day-flying moth which share its habitats and flight period. The larva overwinters fully grown in a cocoon and pupates in the spring without further feeding, resulting in an earlier emergence period than for most Skippers.

Pyrgus malvae (Linnaeus, 1758)   PLATE 2   Grizzled Skipper

Resident; chalk grassland, open woodland; restricted but fairly common.

Univoltine; mid-May to June.

Foodplant – wild strawberry, creeping cinquefoil, [salad burnet, tormentil], (agrimony).

The distribution, and flight period, of the Grizzled Skipper rather closely follows that of the Dingy Skipper, peak populations occurring on the scarp of the North Downs. It

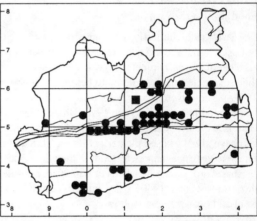

appears on the wing very slightly earlier than the previous species but after the middle of May the two are commonly seen flying together. The Grizzled Skipper is, however, rather less regular in the more open wealden woods than the Dingy Skipper although recorded more often in the Epsom/Ashtead Common area. The only records away from these areas being: Ash 1994 (D. Dell); Witley Common, 1985 (DWB); and Stringers Common, 25.6.88 (GAC). This species has also evidently declined somewhat in Surrey, de Worms (1950) recording it from Ham Common, Wimbledon Common, Esher and Oxshott. Second brood examples are extremely rare, the only record available being Westcott Downs, 9.8.1978 (JP). Larvae were found on the lower leaves of agrimony at Bookham Common, 12.8.1951 (Eagles, *Proc. SLENHS* **1951-52**:16), and at the same site on 10.5.1992 a female was observed ovipositing on tormentil (I. Menzies). Ova have also been found on salad burnet, and ova and larvae observed on many occasions on wild strawberry and creeping cinquefoil (GJ). Recent research has suggested that a range of suitable foodplants is an ecological requirement, small tender leaves being necessary for very young larvae, while the older larvae can eat and spin larger leaves. Overwintering takes place in the pupal stage.

PAPILIONIDAE – Swallowtails

The Papilionidae are a large family of world-wide distribution, only one of which is resident in Britain. Many have the hind-wing produced into a long tail which gives rise to the vernacular name. Larva, of the British species (plate 8), aposematic. One species, long since extinct, has been recorded in Surrey.

Papilio machaon Linnaeus, 1758   PLATES 2,8   **Swallowtail**

ssp. *gorganus* Fruhstorfer, 1922

Extinct resident and vagrant.

Foodplant – (carrot).

The Swallowtail butterfly is currently restricted as a breeding species to the fens and broads of East Anglia where it is represented by the subspecies *britannicus* Seitz. However, in the eighteenth and early nineteenth centuries it was more widespread in southern England and these examples are considered to belong to subspecies *gorganus* Fruhstorfer (Bretherton, 1951a). Records from Surrey for this period are as follows: Windlesham, "a nearly full-fed larva found last week of June 1798, which emerged on 10th August" (Bretherton, 1957); Battersea Fields, "year after year I have been accustomed to find the caterpillars of *machaon*, and have always raised the perfect insect from them, yet, though constantly on the watch, I have never once there detected it in the winged state" (George Austin, *E.W.I.* 1:140). More recent records, some instances of which are considered to be due to migration, especially those in the early 1940s when numbers of the continental subspecies were recorded in Kent and Sussex, are as follows:

1919 – July, two larvae on carrot in the vicinity of **Buckland** (Speyer, *Ent. Rec.* **31**:130).

1934 – **Oxted**, 3.8 (Sheldon, *Ent.* **67**:205).

1935 – **Richmond Park**, 29.6 by Major White (*Ent.* **68**:211).

1941 – **Pitch Hill**, 29.6 (*Ent.* **74**:185).

1943 – Downs near **Guildford**, 18.8 (Turner, *Ent.* **77**:73).

1945 – **Cranleigh**, 23.7 (Hardy, *Ent.* **79**:103); **Guildford**, 27.7 by Kaye and **Haslemere**, 28.7 by Col. Gregson (Riley, *Ent.* **78**:142).

1953 – **Godalming**, 10.6 (*Ent.* **87**:62).

1963 – **Redhill**, 22.6 (Burton, *Ent. Rec.* **75**:207).

1964 – **Selsdon**, fertile female (AS in Evans, 1973).

1982 – **Surrey Docks**, 2.7 (Plant, *L.N.* **65**:21).

1994 – **Redhill**, 7.8 (SWG).

PIERIDAE – White Butterflies

The family Pieridae includes the white and yellow butterflies. Some are resident, some regular migrants, some occasional migrants, and some resident species reinforced by immigration. Hibernation occurs in most species in the pupal stage, although the Brimstone overwinters as an adult; the migrant species are incapable of surviving the winter in this country. The ova are tall and elongate, resembling skittles, and mostly laid singly on the respective foodplants. The larvae (plate 8) are scarcely tapered, unadorned with spines and most feed openly on their foodplants, many being cryptic but some aposematic; those of *Pieris* are frequently pests of cultivated Cruciferae. The pupae are formed fully exposed and head upwards, secured both by a silken pad at the cremaster and by a girdle of silk. Six species are currently resident, with a further one noted as a migrant during the survey period.

Leptidea sinapis (Linnaeus, 1758)   PLATE 2   **Wood White**
ssp. *sinapis* (Linnaeus, 1758)

Resident; woodland, very local but fairly common.

Bivoltine; May and June, and late July and August.

The Wood White is an extremely local species in Surrey, being confined to the wealden woods in the immediate vicinity of Chiddingfold. In these woods it is not uncommon, and as long as the habitat is kept suitably open it would not appear to be endangered. At the turn of the century the Wood White's range had become very

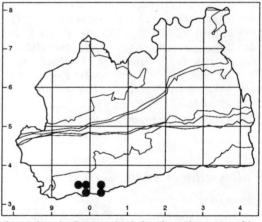

restricted. The *VCH* considered it to be extinct in Surrey, "as it has been in many of its former localities in adjoining counties", although it had been formerly common at Haslemere, and as close to London as Coombe Wood (Stephens, 1828). De Worms (1959), gives a record from Warlingham, 1900, and "a dozen specimens obtained there from 1926-29". Expansion occurred in the middle of the present century, possibly assisted by coniferization of deciduous woodlands (MBGBI 7(1)). Perrins (1959) suggests that it was then to be found in the Cranleigh woods, and gives older records, the date of which is not indicated, from Brook and Witley. The record for Princes Coverts in Evans (1973), although consistent with this theory, is highly suspect and considered erroneous unless the result of an attempted, and undocumented, introduction. The species is bivoltine in Surrey, the second brood, considered by some to be only partial (MBGBI 7(1)), being hardly less numerous than the first.

RECORDS – **Fisherlane Wood**, 13.8.83 (PC); **Tugley Wood**, 1980-82 (TJD), 1985-87 (SCP), 8.6.92, 29.7.92 (GAC), 17.6.93 (PFC), 14.6.94 (G. Revill); **Botany Bay**, 30.5.84, 15.6.85 (PFC), 19.6.88 (RKAM), 1.8.84, 8.6.92 (GAC), 18.8.93 (PFC), 21.8.94 (A. Hoare); **Oaken Wood**, 24.7.94 (M. Bennett); **Alfold**, 20.5.92 (DWB); **Sidney Wood**, 29.7.92 (GAC).

Colias hyale (Linnaeus, 1758)   Pale Clouded Yellow

Migrant; not seen since 1959.

The Pale Clouded Yellow is a migrant species which is absent for long periods but occasionally becomes fairly common. It was apparently more regular in the nineteenth century but has declined in the present one, not being seen in Surrey since 1959. Confusion is likely with the following species, as well as the pale form of the Clouded Yellow (f. *helice*) which is much the commonest of the three, and all examples should be retained for critical examination. Nineteenth century records include: Haslemere, 1868 (Barrett, 1893); Buckland, common in 1876; Surbiton/Chessington area, sparingly in 1893 and commonly there in 1900 (*VCH*). Records this century, which must be considered unconfirmed, are:

1945 – **Leatherhead**, 27.8 (Buckstone, *Ent.* **78**:157).

1946 – **Surbiton**, 11.7; Box Hill, 10.7 (3) (de Worms, ibid..).

1947 – **Juniper Bottom**, July (AAW); **Box Hill** area, 8.8, 7.9 (2), 27.9, 19.10 (Cole in Evans, 1973); **Ottershaw**, 14.9 (Bretherton, 1957); **Walton-on-Thames**, 4.10 (Bretherton, 1957); **Worcester Park**, 6.10 (*Ent.* **80**:291).

1949 – **Walton-on-Thames**, 27.8 (Bretherton, 1957). **Cranleigh** district (*teste* Perrins, 1959).

1956 – **Bookham**, seen flying 9.7 (Tatchell, *Ent. Rec.* **68**:271).

1959 – **Hankley Common**, 2.7 (*Ent.* **95**:175).

Colias alfacariensis Berger, 1948   Berger's Clouded Yellow

Unrecorded.

Berger's Clouded Yellow was only recognized as a species distinct from the Pale Clouded Yellow in 1945. The differences in wing-pattern are subjective, being both subtle and variable, and reliable identification is not possible unless the specimen is caught. Even then it is not always possible to assign the specimen to one species or the other unless the genitalia are checked (see MBGBI 7(1):92). There would appear to be no known example from Surrey, but as a fair number of examples have been recorded from both Kent and Sussex in the past, the possibility should not be excluded if specimens of the species pair are caught in the future.

Colias croceus (Geoffroy in Fourcroy, 1785)   PLATE 3   **Clouded Yellow**

Migrant; irregular.

The Clouded Yellow is a species whose presence in Surrey is accounted for solely by migration as it is unable to survive the winter here. Numbers are very variable; 1983 was a very good year, but in many other years it is entirely absent. Being a migrant species it may occur anywhere in the county, but is most regularly seen on the chalk downs; males may be seen patrolling at high speed searching for females, such habitat being suitable for breeding. Evidence suggests that in 1983 a couple of generations bred locally from the earliest migrants, although the early stages were not recorded. The white form, f. *helice* Hübner, which occurs only in the female, has in the past formed a significant proportion of the population, although in the 1983 immigration percentages were quite low, as indeed were females in general. This form is prone to confusion with the two species of Pale Clouded Yellow; this was certainly the case in 1983 when of many claimed as such most proved to be *helice*. Large migrations were more frequent in the past, notably around the turn of the century and in the 1940s. The decline in its fortunes is probably due to a combination of climatic factors affecting migration and post-war farming practices in Europe.

Gonepteryx rhamni (Linnaeus, 1758)   PLATES 3,8   **Brimstone**

ssp. *rhamni* (Linnaeus, 1758)

Resident; open woodland, commons, heathland; widespread and common.

Univoltine; recorded in every month from February to October.

Foodplant – buckthorn, alder buckthorn.

The Brimstone is a very common and widespread species throughout the county, often being the first species to emerge from hibernation in the early spring. At this time, the

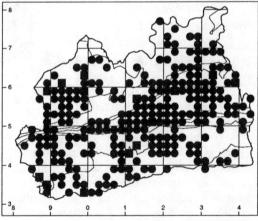

males especially are prone to wander great distances and this explains the distribution being rather more widespread than that of the larval foodplants. The adult is very long-lived; sometimes the last few tattered examples from the spring can be seen flying alongside the freshly emerged examples of the next generation. The larvae are easy to find on the upper surface of a buckthorn leaf, the majority of records being from alder buckthorn, probably because it is easier to search, and the eggs are not too difficult to spot on isolated bushes. Major variation is unknown, but a bilateral gynandromorph was taken at Horsell on 10.8.1958 (Bretherton, 1965).

Aporia crataegi (Linnaeus, 1758)   **Black-veined White**

Extinct.

The Black-veined White was formerly a resident of southern England but has been extinct since at least 1925. Surrey was apparently never a very good county for it and it is not mentioned in Barrett (1893). However, Coombe Wood, seen in plenty in 1810, was given by Stephens (1828), and there is a specimen from Croydon, late 1880s, in the E.B. Ford collection. Hundreds were released on Holmwood Common in the mid-1970s, but none were seen in the following season (Pratt, 1983). Such action, if attempted now, would be in contravention of the Wildlife and Countryside Act, 1981, which quite rightly forbids the introduction of foreign species. Further small scale attempts were stated to have occurred at Chiddingfold Forest *(sic)* in 1984 and Fetcham Downs in 1986 (Oates and Warren, 1990).

Pieris brassicae (Linnaeus, 1758)   PLATE 3   **Large White**

Resident; gardens, commons; widespread and common.

Bivoltine; May and June, and from July to September.

Foodplant – cabbage, honesty, (nasturtium).

The Large White, together with the two smaller members of the genus *Pieris*, are the species known to many people as the Cabbage Whites because of the damage their larvae can do to cultivated members of the

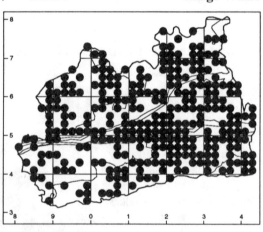

Cruciferae; in fact it is only the Large and Small Whites that do this, the larvae of the Green-veined preferring to feed on wild crucifers. The Large White is very much a butterfly of domestic gardens, allotments and cabbage fields but does fly strongly and is a widespread and often common species in Surrey although the numbers vary considerably from year to year. In part this is due to attacks on the larvae by the hymenopterous parasites of the genus *Apanteles* which can cause a very high rate of larval mortality. It is also likely that migration plays some part in boosting numbers of the resident population – a theory which is difficult to prove in an inland county, but is supported by evidence from neighbouring seaboard counties where considerable swarms of white butterflies are occasionally recorded flying in off the sea. The eggs are laid in batches and the conspicuous larvae feed more or less gregariously, although in my experience, not being a gardener, are seldom seen.

Pieris rapae (Linnaeus, 1758)   PLATE 3   **Small White**

Resident; gardens, commons; widespread and common.

Multivoltine; appearing from late April to September.

Foodplant – cabbage, (nasturtium).

The Small White is another species which is associated with cultivated crucifers and although found throughout the county is only at all common in the immediate vicinity of such crops. In the less agricultural areas of the county it is much less

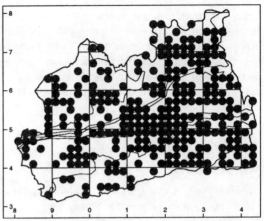

common and certainly less frequent than the Green-veined White with which it may easily be confused. The eggs are laid singly on the foodplant and consequently it is less of a pest species than the Large White; it does however seem less susceptible to parasitic attacks, and possibly also predation, the larva being cryptic rather than aposematic. In common with the Large White it is likely that native populations are reinforced by migration, especially in later broods which are frequently more common. There are several broods, the first being more clearly demarcated than the subsequent ones.

Pieris napi (Linnaeus, 1758)   PLATE 4   **Green-veined White**
ssp. *sabellicae* (Stephens, 1827)

Resident; woodland, gardens, commons; widespread and common.

Bivoltine; late April to June, and July to September.

Foodplant – garlic mustard, (hedge mustard, cabbage, horse-radish).

The Green-veined White, although closely related to the two other *Pieris* species and unfairly lumped with them under the epithet "Cabbage White", is considerably

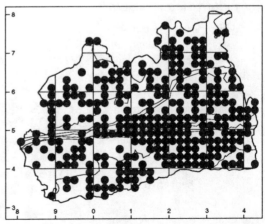

different in a number of ways. The larvae feed mainly on wild members of the Cruciferae, and are not pests of cultivated plants; there are no recent records of the utilization of these plants although cabbage and horse-radish are given as foodplants in Evans (1973). The adult is commonest in damp, shaded areas such as wood edges, lanes and streamsides and has a more even distribution across Surrey than the other two species; it is probably the most frequently encountered white although often misidentified as the Small White, especially males of the spring brood in which the green venation is often indistinct. There is very little evidence of migratory tendencies although the butterfly is a fairly strong flier and capable of local spread. The English form is considered to belong to subspecies *sabellicae* but the treatment of speciation and subspeciation in the *Pieris napi* group is still somewhat controversial.

Pontia daplidice (Linnaeus, 1758)   **Bath White**

Migrant; very rare.

The Bath White is amongst the rarer migrant species to visit us. There are only a handful of nineteenth century records for Surrey, and since 1950 it has been very rarely noticed anywhere in England. During the 1940s it was more regular and in 1945 a mass invasion occurred, centred on the west country but with individuals recorded from all of the southern coastal counties, including seven in Surrey. The initial migration was in July, and the records for Leatherhead at the end of August may well have been the result of local breeding.

19th century – Taken singly at **Headley** and **Box Hill** prior to 1860 (*VCH*);
            Addington, 1894 (de Worms, 1950).
1945 – **Epsom Downs** 15-17.7 (2) (de Worms, 1950); **Tattenham Corner,** 20-23.7 (2)
            (Riley, *Ent.* **78**:143); **Leatherhead,** 27.8 (3) (Buckstone, *Ent.* **78**:157).
1946 – **Cranleigh** district (*teste* Perrins, 1959).

Anthocharis cardamines (Linnaeus, 1758)   PLATE 4   **Orange-tip**
ssp. *britannica* (Verity, 1908)

Resident; gardens, commons,
       open woodland; wide-
       spread and common.
Univoltine; late April to early
       June.
Foodplant – hedge mustard,
       garlic mustard,
       [shepherd's purse],
       (cuckoo flower,
       charlock, field
       pepperwort, horse-
       radish).

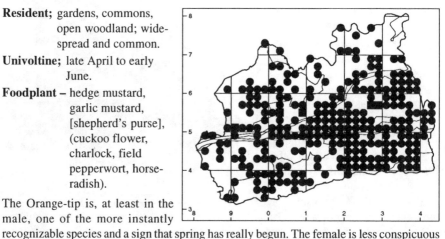

The Orange-tip is, at least in the male, one of the more instantly recognizable species and a sign that spring has really begun. The female is less conspicuous and liable to be confused with the Green-veined or even Small White. The adult prefers more shaded areas and is widespread on the weald and the chalk but less common on the more open heathy areas of the west and only occasional in the more built-up areas of London. The larvae feed on the developing seed pods of a fairly wide range of Cruciferae and the eggs, which are bright orange, and young larvae are easy to find. Larger larvae are more difficult, not least because they are strongly cannibalistic and the several eggs that may be laid on a single plant usually end up as only one larva. The pupae are harder still to find as they are very cryptic and formed away from the foodplant. Second brood examples are exceedingly rare, examples being one at Tulse Hill, 21.8.1952 (Edwards, *Ent. Rec.* **64**:288), and one at Staffhurst Wood, 20.9.1970 (LKE). In particular any apparent female seen after the end of June should be retained as it might be an example of the Bath White (q.v.).

LYCAENIDAE – Hairstreaks, Coppers, and Blues

The family Lycaenidae includes the Hairstreaks, Coppers, and Blues; also included (in accordance with MBGBI **7(1)**) is the subfamily Riodininae, formerly known as the Nemeobiidae. Most species are resident but there are a few, mostly very scarce, migrants. The males are often strikingly different in coloration or pattern to their partners, and have the foreleg somewhat reduced. Most species fly at ground level and are very conspicuous, but the Hairstreak species spend a large amount of their time at tree-top level and so are easy to overlook. When recording these species it is often easier to search for the earlier stages, some species as ova and some by beating the larvae. The ova are mostly flattened and adorned with complex sculpturing although that of the Duke of Burgundy is spherical and smooth. The larvae (plate 9) are woodlouse-shaped, feed on trees or low-growing plants and in several species have a close association with ants; the Silver-studded Blue is entirely dependent on them for larval survival. The subfamily Riodininae, which contains a single British species, differs from the other species by its even more greatly reduced foreleg and by minor wing venational differences; there are considerable similarities in the larval and pupal stages. The various species overwinter as ova, larvae, or pupae, but not in the adult stage. The family as defined comprises thirteen species resident in Surrey, one introduction, and one migrant.

Callophrys rubi (Linnaeus, 1758) PLATE 4 **Green Hairstreak**

Resident; chalk grassland, heathland, open woodland; restricted but fairly common.

Univoltine; May and June.

Foodplant – cross-leaved heath, bird's-foot trefoil, [common rock-rose], (dogwood (berries), gorse, dyer's greenweed).

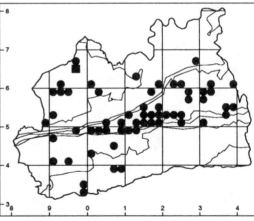

The Green Hairstreak, like the other Hairstreaks, tends to be a species that is overlooked by the casual observer. This is perhaps due more to its relatively smaller size and its coloration which renders it difficult to follow when on the wing and hard to discern when settled, than to the rather skulking habits of the other species. The fairly catholic choice of larval foodplant allows it to occur in a wide range of habitats from woodland rides, heathland and acid grassland to the chalk of the North Downs, and it is in this last habitat that it is undoubtedly the commonest and at its most widespread. The only records from the weald are from the Chiddingfold woods and Canfold Wood, both of which exhibit somewhat heathy characteristics, and it is probably genuinely absent from most of the more typical wealden habitats. Unlike the other Hairstreaks which overwinter in the egg stage, the Green Hairstreak hibernates as a pupa and hence flies rather early in the year, some appearing by mid-April although the peak emergence is in May. The males are territorial and take up a vantage point on a shrub from which they investigate any passing interloper before returning to the same perch. Watching such shrubs on the chalk is the easiest way of finding this species. The white line on the underside, which gives rise to the English name of this subfamily, is, in common with most southerly examples of this species, usually reduced to a few spots, but only very rarely absent altogether.

Thecla betulae (Linnaeus, 1758)   PLATES 4,8   **Brown Hairstreak**

Resident; woodland, wooded commons; very local and uncommon.

Univoltine; late July to early September.

Foodplant – sloe.

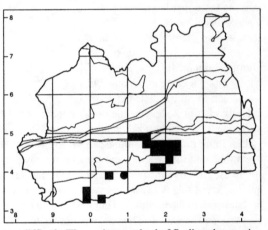

The Brown Hairstreak is a local and very infrequently observed species in Surrey, although, as is the case with most Hairstreaks, it is rather overlooked. With the exception of females engaged in egg-laying the butterflies frequent the upper branches of trees making observation, let alone identification, very difficult. The easiest method of finding the species is to search for the eggs during the winter, the comparatively large and white ova being not inconspicuous on the twigs of sloe, usually in a fork and often on smaller bushes at the edge of a thicket (plate 8). Most of the records have been made in this way, and recent systematic search has revealed it to occur more widely than previously thought. In Surrey, the Brown Hairstreak would appear to be confined to the weald and the southern edge of the chalk, where it inhabits sloe in relatively sheltered positions. No localities for Surrey were given by Newman (1874), nor is Surrey specifically mentioned by Barrett (1893), although he lived fairly close to where it can be found today; South (1906) and Tutt (1907) added our county to Barrett's list but without giving any localities. In addition to wealden records there are a couple of old reports from Ashtead. At a field meeting of the SLENHS at Ashtead Common, 22.6.1901, "many" larvae were beaten (*Proc. SLENHS* **1901**) and further larvae were found on 24.5.1947 (Coulson, *Proc. SLENHS* **1947-48**:61), indicating that the butterfly was probably formerly resident there. A further record from Wimbledon Common in 1986 (Plant, 1987) must be treated with some considerable caution as there is no history of the species from this very well worked site, although it was known from nearby Coombe Wood nearly two hundred years before (Stephens, 1828).

Quercusia quercus (Linnaeus, 1758)   PLATE 5   **Purple Hairstreak**

Resident; woodland, parkland; widespread and common.

Univoltine; July and August.

Foodplant – oak, (Turkey oak).

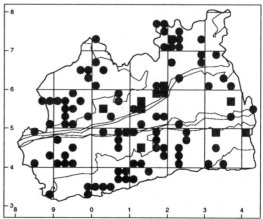

The Purple Hairstreak is a species of woodland and parkland where it inhabits the upper branches of oak trees. Consequently it is not often seen unless specifically searched for, leading some people to consider it an uncommon species, although it is by a long way the most common Hairstreak in Surrey. The butterfly actually occurs widely in the county and is probably to be found in suitable habitat everywhere; certainly it is common on most of the wooded commons in south London as well as more rural areas. Adults can sometimes be seen in swarms around favourite trees, often well into the evening, and have been known to come to moth traps operated below oak trees. The larvae are easy to beat in May, and the ova can be found during the winter on oak buds, but this is hardly the most productive way of finding this species. The larvae probably fall to the ground and pupate within ants' nests – hence the preference for trees in a more open situation (Thomas and Lewington, 1991). The usual adult food source is undoubtedly honeydew on the leaves of oak, but in years of drought, such as 1976, numbers have been observed feeding from damp mud around ponds.

Satyrium w-album (Knoch, 1782)   PLATES 5,9   **White-letter Hairstreak**

Resident; woodland, commons, hedgerows; widespread and fairly common.

Univoltine; late June to August.

Foodplant – wych elm, English elm.

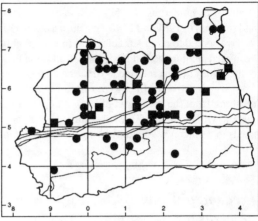

The *VCH* gave several old records of the White-letter Hairstreak from various places in Surrey, including a record of it being found in thousands near Ripley [c. 1810s] by J.F. Stephens. After that it seemed to have become a very rare species until about 1900 when it again appeared in numbers around Ripley, Cobham and Esher. At the latter site, in 1900, some 500 larvae were beaten from a single wych elm. There is, even today, a wych elm in the same area which supports hundreds of larvae. The extensive loss of elms to Dutch elm disease in the 1970s gave considerable cause for concern and indeed at many sites the butterfly was apparently lost. However, like the other Hairstreaks, it is a relatively elusive species and careful observation has revealed it to be present at a number of sites; beating elms in mid-May I have found it at most sites examined and the butterfly is undoubtedly much more widespread than the map indicates, and certainly it cannot be considered endangered. Careful searching would no doubt reveal it to be more widespread in the south-west, and extreme south-east, of the county than the map suggests. In common with most Hairstreaks the adult spends much of its time around the upper branches of its foodplant but occasionally descends to lower levels to feed; hemlock water-dropwort, privet, creeping thistle and bramble are amongst the plants recorded.

Satyrium pruni (Linnaeus, 1758)   PLATE 5   **Black Hairstreak**

Introduction.

Univoltine; June.

The Black Hairstreak is not a resident of Surrey but exists near Cranleigh as an introduction. The original source of this colony was a few individuals which had been reared by A.E. Collier and, being surplus to requirements, were released in his local woods in 1952 (Collier, 1959). From such small beginnings the colony thrived for many years until the wood was ploughed up and converted into arable farmland. The butterfly was considered to have been lost until it was rediscovered fifteen years later in a woodland about a mile from the original site (Thomas and Lewington, 1991). Five separate colonies "developed" in this area which then held the largest numbers of any wood in England (ibid.). I believe that these colonies resulted from further undocumented introductions by another individual rather than by natural spread; the butterfly is dependent on mature stands of sloe, thought to be due to its reluctance to spread to younger, developing stands of the foodplant. It was recorded as numerous on 25.6.1980 in an area about to be destroyed by re-afforestation and searches in 1981 and 1982 were unsuccessful (RFB). More recently it had been reported to have spread to other suitable areas in that vicinity, although I understand that it has not been seen at all in the last few years. In addition to this colony an article on the lepidoptera of the Witley district referred to larvae being found on sloe within twelve miles of Haslemere from which six butterflies were reared (Talbot, *Ent. Rec.* **32**:40). This was a serious error, resulting from the confusion of bought pupae of the Black Hairstreak with wild stock of the Brown Hairstreak (Hawker-Smith, *Ent. Rec.* **36**:75), and demonstrates the care which needs to be taken when considering, in particular, older records.

Lycaena phlaeas (Linnaeus, 1761)   PLATE 5   Small Copper
ssp. *eleus* (Fabricius, 1798)

Resident; grassland, heathland; widespread and fairly common.

Multivoltine; two or three broods from May to September.

Foodplant – (common sorrel).

The Small Copper is a widespread butterfly in Surrey occurring without any apparent preference for habitat. It is equally at home on the chalk downs as on the heaths, and on the commons of south London as in sunny woodland rides. Females of

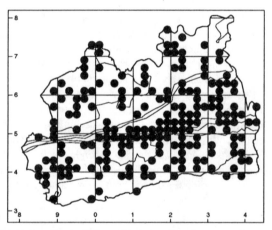

the summer generation select only plants growing in full sunshine to oviposit (MBGBI 7(1)), and thus it would seem that the openness of the habitat is the principal ecological requirement. Overwintering larvae hibernate in any one of the first three instars (ibid..), which results in a protracted emergence of the first brood. In addition there can be two, three or very rarely four broods in a year which to some degree overlap enabling the adult to be seen in almost any month from May to September or October. The adults can be seen at flowers and the males are territorial, chasing away any butterflies that approach them. The extremely rare ab. *schmidtii* Gerhard was seen in Surrey in 1974 (Willmott, *Ent. Rec.* **87**:55).

Lampides boeticus (Linnaeus, 1767)   PLATE 6   Long-tailed Blue

Migrant and probable artificial introduction.

September and October.

Foodplant – (everlasting pea).

The Long-tailed Blue is a widespread tropical species which occasionally migrates as far north as Britain where it is usually seen in the late summer. It is continuously brooded and quite incapable of surviving the British winter although evidence from the 1952 sighting suggests that it might be able to produce a generation during the summer months. At Ranmore, 6.7.1952, a female was caught flying about a clump of everlasting pea, and, once its identity had been established, a search of the plant revealed 26 ova. From these ova a series of specimens were bred in August of that year. When the larvae were nearly fully grown, the plant was removed and confined indoors and on the 22nd of August a further specimen emerged, suggesting that the species was capable of completing its life-cycle under natural conditions (Chevallier, 1952). Most of the records probably relate to genuine migrants, but the possibility of artificial introduction cannot be discounted particularly as larvae are intercepted in most years in various imported beans (MBGBI 7(1)).

1945 – **Epsom Station**, 27.8 (Buckstone, *Ent.* **78**:157).

1949 – **Streatham**, 7.9 (de Worms, 1950).

1950 – **Wimbledon**, 13.10 (Craske, *Ent.* **84**:47).

1952 – **Ranmore**, 6.7. (Chevallier, 1952).

1959 – Near **Guildford**, 14.10 (Johnson, *Ent. Rec.* **71**:267).

1971 – **Wisley** RHS gardens, a female, 4.9 (RFB).

1977 – **Surbiton**, seen in a cemetery (Smith, *L.N.* **57**:84).

1979 – **Sutton**, 15.9, in a house (Brown, *Ent. Rec.* **92**:95).

1980 – **Surrey Docks**, 26.8 (Murdoch, *L.N.* **60**:104).

1990 – **Kew Gardens**, 2.9. (R. Hastings).

Cupido minimus (Fuessly, 1775)   PLATE 6   **Small Blue**

Resident; chalk grassland; restricted but fairly common.

Bivoltine; late May and June, and late July and August.

Foodplant – kidney vetch.

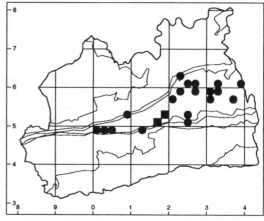

The Small Blue is a species of chalky soils in Surrey and is consequently fairly widespread in the east of the county. Unlike the other Blues which tend to be at their commonest on chalk downland, the foodplant's preference for broken or disturbed ground means that the Small Blue is as often found in such localities as disused quarries and commons as on the prime downland of the scarp slope of the North Downs. The main brood appears at the end of May and throughout June, and in some colonies there is a partial second brood. At Banstead Downs this occurs every year (DC, pers. com.), possibly as a result of higher average temperatures close to the metropolis. Several recent records have been made by searching for ova on the foodplant – a useful method in poor weather when the adults may be difficult to locate.

RECORDS – **Pewley Downs**, 22.5.82 (ASW), 4-15.6.85 (IDMacF), 15.7.94 larvae (M. Bennett); **Merrow Downs**, 1.6.87 (DWB), 1991 (E. Wood); **Newlands Corner**, 17.6.94 (GJ); **Sheepleas**, 17.6.94 (GJ); **Westcott Downs**, 6.83 (JP); **Box Hill**, 23.7.84 (PFC), 27.6.94 ova (GJ); **Headley Warren**, 3.6.92, 5.6.93 (HM-P), 19.7.94 larvae (GJ); **Colley Hill**, 23.6.83 (BC), 27.6.87 (JBS/PFC); **Epsom Downs**, 22.6.88 (GAC), 1992, 93 (ME),12.6.94 (G. Revill); **Howell Hill**, 13.6.94 (GJ); **Banstead Station**, 10.6.84 (R. Hastings (per CWP)); **Banstead Downs**, 25.7.80 larvae on kidney vetch (PJH), 16.6.85 (PFC), 29.5.85, 16.8.86, 29.5.89 (AAW), 27.5.90 (RDH), 14.6.91, 8.6.92 (GAC); **Park Downs**, 8.6.83 (PFC); **Happy Valley**, 27.6.84 (PFC); **Dollypers Hill**, 1992 (P. Grove); **Riddlesdown**, 3.7.85, 24.6.86 (GAC); **Riddlesdown Quarry**, 16.6.85, 10.6.87, 28.5.90 (GAC), 2.7.90 (JC); **Addington**, 14.6.82 (TJD), 19.6.85, 5.7.91 (GAC), 26.6.94 (PFC); **Woldingham**, 18.6.94 (GJ).

Plebejus argus (Linnaeus, 1758)   PLATE 6   **Silver-studded Blue**

ssp. *argus* (Linnaeus, 1758)

Resident; heathland; restricted but common.

Univoltine; late June to early August.

Foodplant – ([heather]).

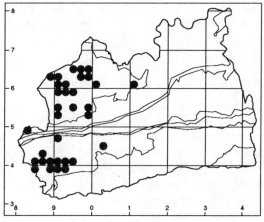

The nominate form of the Silver-studded Blue is a species of heathland and is to be found on all the major heaths in the west of the county, although it does tend to be colonial and one may have to search for the centre of population. On the more isolated remnants of heathland further east it is still found on Blackheath and as close to London as Fairmile Common, although it seems to have been lost from Wisley Common and Oxshott Heath where it still occurred less than fifty years ago (de Worms, 1950). At the present time it is widespread and frequently very common on suitable habitat and the only threat would seem to be loss of heathland to housing development or changes in the nature of the habitat such as grass and bracken invasion following accidental and uncontrolled fires. The larva of the Silver-studded Blue has a very close relationship with various ant species; the females only lay eggs where they can detect pheromones of ants of the genus *Lasius*, and both larvae and pupae live in ants' nests. On a field meeting of the SLENHS to Chobham Common, 18.6.1955, a larva "which would appear to be that of *Plebejus argus*" was found under a log inhabited by the ant *Lasius niger* (Proc. SLENHS **1955**:75). Apart from this single example, the only other instance of the larva being found is the one illustrated in South (1906) which was found crawling under heather on the last day of May at Oxshott, and successfully reared on this pabulum.

ssp. *cretaceus* Tutt, 1909

Extinct.

The Silver-studded Blue in Britain is considered by some to comprise four subspecies, although those other than the nominate one are of doubtful status (MBGBI **7(1)**). The taxon *cretaceus* Tutt is applied to those forms occurring on chalk or limestone and whose larvae feed principally on bird's-foot trefoil and horse-shoe vetch. Specimens of this form differ from the nominate subspecies in their larger size, and the brighter blue coloration of the males. With the exception of some colonies in Dorset, which are far less extreme than were the insects from south-east England, this form is considered to be extinct. The former range of *cretaceus* included the downs of Kent, and east Surrey where it occurred near Addington. An account of this colony can be found in Evans (1973); specimens were last seen in 1957, by which time the site had almost been lost due to scrub invasion, and a few years later it was ploughed and the habitat completely lost. In Kent it survived slightly longer but had virtually gone by 1966. Of the form in Kent, it was stated at a SLENHS meeting in 1919 that the butterfly had all but disappeared three times in the course of the previous 30 years, so the populations evidently fluctuated greatly, the final extinction being probably due to scrub invasion as a sequel to myxomatosis producing conditions unsuitable for its ant hosts..

Aricia agestis ([Denis and Schiffermüller], 1775)   PLATE 7   **Brown Argus**

Resident; chalk grassland, heathland; restricted but fairly common.

Bivoltine; late May and June, and late July and August.

Foodplant – [common rock-rose, cut-leaved crane's-bill].

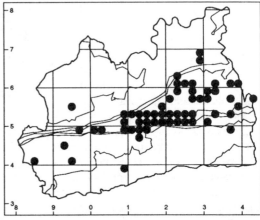

The Brown Argus is a species of open habitat which is fairly common over the whole of the chalk but rarely recorded elsewhere. Indeed the only records away from the chalk are: Pirbright, 14.8.94 (GAC); Frensham Common, 17.8.88 (GAC); Shackleford, 1994 (J. Ruffin); Milford, 1984 (DWB); Ewhurst, 1991 (SFI); South Godstone, 27.8.80 (RDH); and Mitcham Common, 10.7.90, 18.6.91 (DCL), 20.8.94 (A. Shelley). At this latter site is has only recently appeared, not being noticed by earlier recorders, and is considered by the recorder (DCL) to be associated with cut-leaved crane's-bill. The heathland records refer to single individuals but the butterfly is probably rather more common in such areas than records suggest; the Brown Argus is easily confused with female examples of the Silver-studded Blue or Common Blue unless examined closely or caught. In these areas it is probably associated with plants of the geranium family, but on the chalk the generally accepted foodplant, common rock-rose, is certainly used; females have been observed laying on it on many occasions (GJ).

Polyommatus icarus (Rottemburg, 1775)   PLATE 7   **Common Blue**

ssp. *icarus* (Rottemburg, 1775)

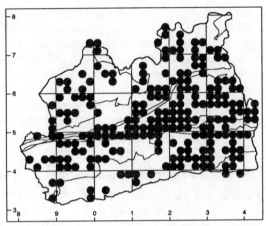

Resident; chalk grassland, commons, heathland, woodland rides; widespread and common.

Bivoltine; late May and June, and late July to early September.

Foodplant – bird's-foot trefoil, [black medick, tall melilot].

The Common Blue is the commonest and most widespread member of its subfamily, being found in open areas throughout the county. It is probably commonest on areas of chalk grassland, but still frequent on commons on the clay and on heathland. To the more casual observer the male resembles the Adonis Blue but is less brightly coloured and lacks the marginal black and white chequering of that species, and the female could be mistaken for the Brown Argus. Completely brown female Common Blues are rare in Surrey, most having considerable amounts of blue on the basal area of the wings, but the differences in underside spotting should resolve any doubts. The larvae feed on plants of the family Leguminosae, and have been recorded from bird's-foot trefoil, with ovipositing noted on the other plants listed.

Lysandra coridon (Poda, 1761)   PLATE 7   **Chalk-hill Blue**

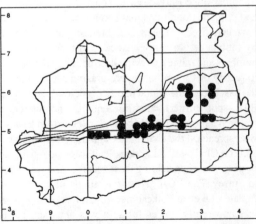

Resident; chalk grassland; restricted but fairly common.

Univoltine; mid-July to early September.

The Chalk-hill Blue, as its name suggests, is a butterfly of calcareous soils and in Surrey is restricted to the chalk. Within this formation it is not uncommon both along the scarp of the North Downs and on many areas of chalk grassland on the dip slope. Colonies can only survive in areas where horse-shoe vetch grows; other members of the Leguminosae can be utilized occasionally allowing colonization of non-calcareous areas, but these colonies soon die out. The number of adults in most colonies has undoubtedly decreased since the middle of this century, probably as a result of scrub encroachment into its breeding areas. At Riddlesdown this was attributed to withdrawal of steam engines on the nearby railway, which had caused periodic fires which kept the scrub down; in other areas the advent of myxomatosis also resulted in loss of suitable habitat. The adult is single-brooded and usually appears between the two broods of the Adonis Blue, although some adults persist into late August and early September thus overlapping with the second brood of the latter species. The females of the two species can be extremely similar leading to misidentification. The larvae are also similar, but are found at slightly different times, that of the Adonis Blue overwintering as a small larva whilst the Chalk-hill Blue passes the winter as an egg. It is also stated that the larva of this species feeds at dusk and by night whereas the larva of the Adonis Blue feeds by day.

Lysandra bellargus (Rottemburg, 1775)   PLATE 7   **Adonis Blue**

Resident; chalk downland; very local but fairly common.

Bivoltine; June, and August.

Foodplant – (horse-shoe vetch).

The Adonis Blue is currently a very local species in Surrey and is confined to the better stretches of chalk downland between Guildford and Reigate. There is currently serious concern about this species as it has not been seen west of Ranmore since 1991, despite habitat and climatic conditions that should

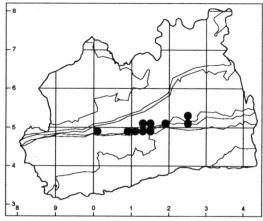

have benefited it. The essential requirement of the habitat is short-cropped turf maintained by rabbits or grazing livestock, and there would appear to have been a decline in both numbers and range following destruction of the rabbit population by myxomatosis. Attempts to rectify this by rather severe grazing regimes on some areas of the North Downs has in the past been detrimental to other species of lepidoptera such as *Zygaena trifolii palustrella* (Five-spot Burnet) which are equally endangered. Reduction in grazing levels has now restored a more equitable balance. Records in Evans (1973), suggest that it was once more widespread on the chalk, although the great similarity between the females of this species and the Chalk-hill Blue does lead to misidentification and considerable caution needs to be used when considering these records. The Adonis Blue has probably always been resident in Surrey. The *VCH* gives an example of an early attempt at introduction; several fertile females, taken at Folkestone, were released on Reigate Hill in 1876 by Sydney Webb. This was apparently successful as it was abundant in June 1900, although the possibility of local spread from further along the downs cannot be ruled out, especially as Newman, writing in 1874, gave it as abundant at Mickleham, also at Guildford and on the south side of the Hog's Back. It was also introduced to Pewley Down [?1970s] (Oates and Warren, 1990), but has not been noted there since 1987.

RECORDS – **Pewley Down**, 4.6.82, abundant (DWB), 4-15.6.85 abundant, 1987 (IDMacF); **Hackhurst Downs**, 1991 (J. Cranham); **White Downs**, 15.6.85 (JC); **Westcott Downs**, 29.8.83 (RDH), 29.5.85, 10.8.90, 14.6.91, 22.5.92, 2.9.94 (GAC); **Ranmore**, 1980-85 (RFB), 29.6.80, 19.6.82 (TJD), 5.6.81, 4.6.84 (GAC), 10.6.85 (PFC); 13-27.8.83, 31.5.84, 19.8.84, 31.5.85 (PC), 30.5.92 (GJ); **Box Hill**, 1992, 93 (A. Reid); **Colley Hill**, 16.6.84 (BC), 21.6.86 (JBS).

Cyaniris semiargus (Rottemburg, 1775)   **Mazarine Blue**

Extinct.

The Mazarine Blue is a species whose history in Britain is subject to considerable doubt and confusion (Bretherton, 1951b, and Allan, 1980) – the former author suggesting that there were fewer properly authenticated British examples of it than of the Large Copper. It is probable that it was once resident and certain that it is now extinct; its demise however is clouded by the probability of fraudulent examples and some doubt must attach to many of the examples reported from Surrey. Our county is amongst those mentioned by Stephens (1828), the locality being Windlesham Heath "towards the end of May and of July", which Dale (*E.M.M.* **38**:78) gives in greater detail as "captured July 16th, 1878, by the Rev Dr Abbott". Of this record Tutt (1908) says "should no doubt be 1818". These authors imply that a single specimen was involved whereas the punctuation in Stephens suggests that the dates given refer to this single site rather than the several other ones listed. The *VCH* has it as being reported to have been taken on Reigate Hill and near Headley [prior to 1870]. A specimen, confirmed by the British Museum, was amongst a series of blues collected at Croham Hurst in August 1880 (Olliff, *Ent.* **14**:43). Bretherton (1951b) adds: Riddlesdown, 5.9.1877 (Standish: Dale register), and "a rather doubtful pair" at Mickleham in 1904 or 1905. The original reference, *Ent.* **39**:24, states that a pair were taken by Beattie or his daughter during 1904 or 1905 along with *H. paniscus* (i.e. Chequered Skipper) – ample reason for Bretherton's doubt. Nearly all of these records seem to be based on hearsay rather than hard evidence, and it is perhaps significant that none of the better known entomologists of that time appear to have seen the butterfly in the wild.

Celastrina argiolus (Linnaeus, 1758) PLATES 10,9 **Holly Blue**
ssp. *britanna* (Verity, 1919)

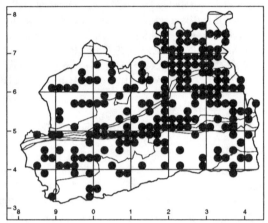

Resident; woodland, commons, gardens; widespread and fairly common.

Bivoltine; May and June, and mid-July to August.

Foodplant – ivy, dogwood, firethorn, (holly).

The Holly Blue is a species that is subject to a regular short term cycle of decline and expansion, probably due to interaction with its host-specific parasite, the ichneumon *Listrodomus nycthemerus*. Currently it is on the decline from one of its most prolific expansions of recent years in which it appeared in numbers sufficient to attract interest from the general public. During this period of abundance large numbers were reported throughout the county, but in leaner years it seems to favour particularly the parks and larger gardens of south London. The adult butterfly has a characteristic flight, completely unlike that of most blues, in which it flies high along the edges or tops of banks of trees and shrubs. The Holly Blue usually has two broods, larvae from the first feeding on holly in May and June and larvae of the second utilizing ivy in September, the preference being for the flowers and berries rather than leaves. From its often suburban distribution it can be inferred that other foodplants can be used, especially by the first brood larvae, and recently ova were noticed being laid on, and the larvae observed feeding on the berries of, both dogwood and firethorn. The winter is passed as a pupa and the butterfly is usually the first Blue to be seen during the year. A bilateral gynandromorph was taken at Chessington, 23.9.1987 (AMJ).

Hameuris lucina (Linnaeus, 1758) PLATES 10,9 **Duke of Burgundy**

Resident; chalk grassland; very local and scarce.

Univoltine; mid-May to early June.

Foodplant – cowslip.

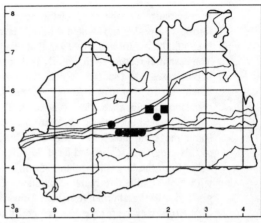

The Duke of Burgundy is currently a very local species in Surrey, its decline and change in habitat preference mirroring its national status. At the present time it is restricted to chalk grassland in areas along the scarp of the North Downs west of Dorking and on either side of the Mole valley north of the Dorking gap. The colonies are very small and easily overlooked; it is also likely that some shown on the map are no longer extant. Ecological requirements are for a certain amount of scrub, around which the males establish territories and which may also shelter the foodplants from draughts. Blanket scrub removal and overgrazing of the downs are probably detrimental to this species. During this century there has been a change from woodland habitats to its present grassland ones, this being attributed to reduction in coppicing of the woodland followed by scrubbing of the downland after myxomatosis (MBGBI 7(1)). The ova may be found on the underside of the leaves of the foodplant in early June (plate 9), but larvae are more difficult to locate as they feed by night and hide in leaf litter during the day. It was evidently once more widely distributed; Barrett (1893) says "probably to be found in all large woods in the southern half of England". This may have been an exaggeration. The only other records from localities not presently occupied are: Newlands Corner, before the war and 1954 (3), also at Durfold (Perrins, 1959); Canfold Woods 1950 and 51 "but not occurring there regularly" (*teste* Perrins, 1959); Bookham Common, a few seen (Wheeler, 1955); and Hog Wood, 10.5.1953 (*Proc. SLENHS* **1953-54**). This latter site straddles the Surrey/West Sussex border and it is not clear to which vice-county the record refers.

RECORDS – **Clandon Downs**, 2.6.92 (J. Pontin); **Combe Bottom**, 1982, 83 (JC); **Hackhurst Downs**, 6.83 (6) (M. Tickner (per DCL)), 1989 (J. Cranham), 28.5.90 (A. Lickley); **White Downs**, 6.82 ova on cowslip (MG-P), 10.6.83 ova on cowslip (PMS), 7.5.88 (AJH), 27.5.91 (T. Smithers), 16.5.92 (K. Reel); **Westcott Downs**, 5.92 (JP), 21.5.92 (P. Churchill); **Norbury Park**, 17.5.80 (RAC); **Fetcham Downs**, 29.5.81 ova (BC); **Headley Warren**, 8.6.86 (PFC), 25.5.91, 11.6.94 ova (GJ).

NYMPHALIDAE – Tortoiseshells, Fritillaries and Browns

In accordance with the nomenclature adopted in MBGBI 7(1), the former families Satyridae and Danaidae have been reduced to subfamily status. It should be mentioned, however, that this is not universally accepted. The essential feature of the family is that the foreleg is much reduced in both sexes. Most species are residents, but the subfamily Nymphalinae contains a number of well known migrants, some regular and some of occasional appearance. The larvae of the Tortoiseshells, Fritillaries and closely related species (plate 9) are usually heavily spined and feed, sometimes communally, on trees, common nettles and violets. The pupae of this group are suspended head downwards by the cremaster, often away from the larval foodplant, and in some species are ornamented with highly reflective metallic patches, of silver or gold hue, which gives rise to the alternative name of chrysalides. The Satyrine larvae (plate 9), however, are tapered, have a bifid anal segment, and are cryptic being coloured green or brown, some with longitudinal stripes. They feed on grasses and overwinter in this stage. Several members of this subfamily scatter their eggs when in flight rather than selecting a particular plant on which to oviposit. Seventeen species are resident in Surrey, with a further two occurring as regular migrants.

Ladoga camilla (Linnaeus, 1764)   PLATE 10   White Admiral

Resident; woodland, wooded commons; fairly widespread and fairly common.

Univoltine; late June to early August.

Foodplant – honeysuckle.

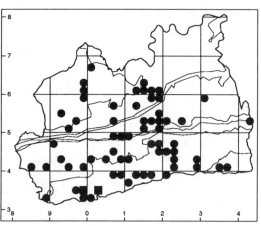

The White Admiral is currently a fairly widespread species in Surrey although more or less absent from the north-east of the county. It was not always so frequent and Barrett (1893) felt that it had totally disappeared from some of its haunts in the south-east in recent years, and the *VCH* gives only one record of a few specimens being taken in a wood near Horsley in 1901, adding that as there had been no other records from Surrey it may have been deliberately introduced. In the early years of this century it increased rapidly, reaching Wimbledon by 22.7.1917 (Wakely, *Ent. Rec.* **30**:154). After this period of expansion the population became more stable, and apart from small declines in the north-east it is probably as common now as at any time since the *VCH* was published. The adult occurs in wooded areas and favours its foodplant growing in moderate shade to breed. The larva hibernates in a folded leaf attached to the stem by silk and has been found in this stage. Aberration is restricted to the loss of part or all of the white band and ab. *nigrina* Weymer has been recorded from Oxshott, 1993 (JP), and was, in addition to ab. *obliterae* Robson and Gardner, observed at Bookham Common in 1994 (Willmott, *L.N.***73**). A probably unique bilateral gynandromorph was taken near Dorking on 16.7.1986 (Beccaloni, *Ent. Rec.* **100**:102).

Apatura iris (Linnaeus, 1758)   PLATE 11   Purple Emperor

Resident; woodland; restricted but not uncommon.

Univoltine; July and early August.

Foodplant – (goat willow).

The Purple Emperor, despite being the largest species of butterfly found in Surrey, is also one of the most overlooked, leading many people to consider it to be rare. Its treetop habits and the difficulty of observing it above the canopy of dense woodland mean that it is only seen on occasions when it descends

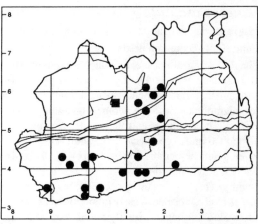

to lower levels to feed and in particular to oviposit. The map shows the Purple Emperor to be found throughout much of central and south-western Surrey although it is more restricted in the eastern weald. Within this range it probably occurs in most of the larger tracts of woodland and cannot by any stretch of the imagination be considered endangered. In 1994 there was a confirmed record from Kenley Common, 16.7.94 (M. Enfield), and an unconfirmed one from nearby Dollypers Hill; some doubt must attach to the origin of these insects, and they are not shown on the map. Edward Newman recorded the Purple Emperor as being formerly abundant near Godalming, but by the turn of this century the *VCH* considered that it had probably disappeared from Surrey. Even Barrett had seen only two examples in Surrey, in a wood near Haslemere. The *VCH* also says "a tradition exists that the species occurred in the early part of the last century [19th] in Prince's Covers". It was recorded from here in August, 1943 (de Worms, 1950) and can still be found there today (see records).

RECORDS – **Gibbet Hill**, 28.7.86 (DWB); **Haslemere** (SHC);
Fisherlane Wood, 1982 (DWB), 17.7.86 (SCP); **Tugley Wood**, 11-17.7.82 (TJD), 18.7.85 (4), 6.8.85 (6), 9.7.87 (SCP), 18.7.94 (PFC); **Botany Bay**, 1.8.84 (GAC), 21.7.94 (D. Dell); **Oaken Wood**, 11.7.94 (P. Farrant); **Pockford Bridge**, 7.8.87 (RKAM); **Milford**, 11.7.94 (DWB); **Royal Common**, 17.7.91 (J. Pontin); **Winkworth Arboretum**, 3.7.82 (RFB); **Bramley**, 7.83 (RFB);

Sidney Wood, 16.7.85 (IDMacF); **Ewhurst**, 1988 (SFI); **Elm Corner**, 22.6.87 larva (per AJH); **Bookham Common**, 14.7.85 (D. A. Boyd (per CWP)), 15.7.89, 18.7.92, 30.6.93 (G. Revill), 23.7.94 (K. Reel); **Leith Hill**, 1980 (SBRC); **Wallis Wood**, 21.7.94 (GJ); **Vann Lake**, 20.7.88 (G. Blaker), 17.7.94 (GJ); **Norbury Park**, 1984-86 (SBRC); **Princes Coverts**, 25.7.82, 29.6.92 (JP); **Holmwood Common**, 1.8.84 (c.4) (GAC); **Ashtead Common**, 30.7.84 (c.6) (GAC), 25.7.87 (c.6) (GG), 15.7.92 (PFC), 8.7.93 (GJ), 20.7.94 (B. Hilling); **Epsom Common**, 15.7.86 (Mrs P. Dawe (per CWP)); **Headley Heath**, 31.7.81 (J. Bebbington), 7.7.93 (A. Reid); **Box Hill**, 23.7.94 (A. Kerr); **Juniper Top**, 27.6.92 (P. Churchill); **Glovers Wood**, 26.6.92 (RDH).

Vanessa atalanta (Linnaeus, 1758)   PLATE 11   Red Admiral

Migrant; annual and often fairly common.

Foodplant – common nettle.

The Red Admiral is a species which is migratory to Surrey, originating from North Africa and southern Europe. During the survey period, 1980-94, it was recorded in every year although the actual numbers fluctuated widely, no doubt due to climatic conditions suitable for migration and varying numbers breeding around the Mediterranean. The Red Admiral certainly breeds here, and the larvae, feeding singly and amongst spun leaves, are not difficult to find in good years. The adult is usually seen in the late spring, but is commonest in the autumn as numbers are boosted by locally bred examples and possibly further migrants. There is still some controversy over whether it is capable of successful hibernation in this country. This may occur on occasion, but the presence of examples in March and April is just as likely to be due to early immigration as to local overwintering; there is no doubt that survival in this country is only possible by continual migration. The adults are often seen in gardens where, in addition to feeding on flowers such as buddleia and Michaelmas daisies, they are also to be found on ivy bloom and rotting fruit. On heathland in west Surrey the adults were observed being attracted to fermenting sap arising from birch trees infested with larvae of *Cossus cossus* (Goat Moth).

Cynthia cardui (Linnaeus, 1758)   PLATE 11   Painted Lady

Migrant; annual and often fairly common.

Foodplant – thistle, (spear thistle, viper's bugloss).

Like the Red Admiral, the Painted Lady is a migrant species of Mediterranean origin. It too is seen virtually every year, although during the survey period there were no dated records for 1981 or 1984. The numbers involved are also frequently much smaller than those of the Red Admiral, and bumper years are uncommon. The larvae are seldom observed, ironically the only record coming from 1989, in which year there was only one record of the adult received. There is no evidence that successful overwintering occurs, or that it is even attempted, the butterfly being continuously brooded with no diapause stage. With both of these migrant species there is some evidence that a return migration can occur in the autumn, a phenomenon only observable in the southern coastal counties, but, in this species at least, unlikely to have much effect on reinforcing numbers of the Mediterranean population.

Aglais urticae (Linnaeus, 1758)   PLATE 11   **Small Tortoiseshell**

Resident; commons, gardens, woodland rides; widespread and common.

Bivoltine; peaks in April, July, and September.

Foodplant – common nettle.

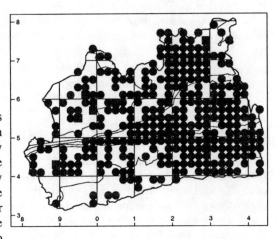

The Small Tortoiseshell is a species of almost ubiquitous distribution in the British Isles, and is similarly very widely distributed in Surrey. The adults are prone to wander and may occur in sites which are unsuitable for breeding, which makes their apparent absence from the extreme south-west of the county difficult to explain. It is certainly partly a result of under-recording from this area, but may also be connected with soil type and lack of larval foodplant. Largest numbers occur in agricultural areas, especially in the weald, where common nettles benefit from the nitrate-enriched soil. Numbers tend to vary from year to year, and are probably reinforced by migration. In 1991 post-hibernation adults were very scarce and in the spring larval nests, usually abundant, were virtually non-existent, but in the summer large numbers of butterflies suddenly appeared and very likely had a continental origin. There are probably two broods in most years, but it has been suggested that some adults of the first brood will hibernate rather than producing a second – a situation not unlike that found in the Comma (q.v.). Adults fly late into the autumn, often being seen in October and November, before hibernation, and will often fly in mild periods in late winter. The natural hibernation site would be in caves and hollow trees, but overwintering adults are most often seen in wartime pill-boxes, garden sheds and even in houses.

Nymphalis polychloros (Linnaeus, 1758)   PLATE 12   **Large Tortoiseshell**

Status uncertain; probably only temporary resident this century.

Foodplant – (sallow, poplar).

The Large Tortoiseshell is one of our more enigmatic species. The records from the previous century are not precise enough to determine the exact status of the species in Surrey, but it was clearly fairly common at times, existing just into the present century. It then remained unrecorded until 1917 when again it was not uncommon for a period of four years, and certainly bred. Apart from a single record in 1930, it again disappeared until the 1940s, and in particular the period 1947-50, when there were many sightings, although not of the early stages. A further rapid decline followed and to date only a few further records have been made, although the records from Chobham, 1972-3, suggest that a brood had been raised. Thus, whilst it may have been resident in the nineteenth century, it has never occurred for more than a few years at a time in this century, and should properly be considered as a temporary resident. Various factors have been suggested for this apparent change in status including host-parasite interactions and climatic change. The picture is further confused by a number of misguided individuals who release bred, continental examples of this butterfly.

19th century – Barrett (1893) gives – "at one time very common immediately round London and in many other places from which it has now partly or wholly disappeared". The *VCH* considered it to be generally distributed throughout the county listing: **Haslemere, Chiddingfold, Cranleigh, West Horsley, Leatherhead, Esher Common, Claygate, Kingston** and **Worcester Park. Ranmore Common,** seen frequently in the 1880s and 90s (Buckstone, Ent. **81**:271).

1901 – **Streatham Station** (Plant, *L.N.* **65**:21).

1902 – **Reigate,** 24.5, and **Ranmore Common,** 7.6(Carr, *Proc. SLENHS* **1902**).

1917 – Several at **Horsley** (*Proc. SLENHS* **1917-18**).

1918 – **Wormley,** "not uncommon, a series bred from larvae on poplar" (Tullett, *Ent. Rec.* **32**:52); **Kingston,** 16.5 (Ing, *Ent.* **51**:161).

1919 – Spring, a number seen in Surrey (*Proc. SLENHS* **1919-20**); **Ranmore Common,** 3 on sallow blossom (Buckstone, *Ent.* **81**:271).

1920 – **Ashtead Woods,** 21.3 (Barnett, *Proc. SLENHS* **1920-21**).

1930 – **Tatsfield,** August (Barnett, *Proc. SLENHS* **1930-31**).

1940 – One seen by Mr. Palmer in September at **Thames Ditton,** and another a few weeks earlier at **Molesey** (de Worms, *Ent.* **74**:43).

1942 – **Chipstead,** 22.6 (Frohawk, *Ent.* **75**:171).

1945 – **Ewell,** 5.7 (Buckstone, *Ent.* **79**:90), **Hambledon,** 5.7 (Talbot, *Ent.* **78**:125).

1947 – **Worplesdon Hill,** 12.4 (Howell, *Ent.* **80**:164); **Claygate,** 13.4, 26.5, and 2.6, possibly the same example (Stallwood, *Ent.* **80**:260).

1948 – **Box Hill,** March (*Proc. SLENHS* **1948-49**:4); **Friday Street,** 13.3 (ibid.. **1948- 49**:11); **Bookham Common,** 27.3 (Wheeler, 1955); **Ashtead Woods,** 7.4 (2) (Buckstone, *Ent.* **81**:271); **Ranmore,** 7.5 (Buckstone, *loc. cit.*); **Fisherlane Wood,** 9.5 (RFB); **Mickleham Downs,** 17.7 (*Proc. SLENHS* **1948-49**:73); **West Croydon,** 23.7 (EHW); **Westcott,** 27.7 (Buckstone, *Ent.* **81**:271); **West End, Esher,** August (Palmer, *Ent.* **82**:166); **Blackheath** (*teste* Perrins, 1959).

1949 – **West End, Esher,** 26.3 (Palmer, *Ent.* **82**:166); **Betchworth,** 9.4 (G.A. Cole); **Weybridge Station,** April 1949 (JLM).

1950 – **Effingham,** 11.4 (CGMdeW); **Chiddingfold,** 22.5 (*Proc. SLENHS* **1950-51**:11).

1952 – **Givons Grove,** 1952 (JDH).

1953 – **Barnsthorns Wood,** 18.4 (*Proc. SLENHS* **1953-54**:5).

1964 – **Godalming,** 19.7 (Williams, *Proc. SLENHS* **1967**:72).

1972/3 – **Chobham, Longcross,** 1972, one seen; 18.4.73, one caught, about six seen (per CGMdeW).

1980 – Near **Guildford,** 13.9 (Miss J. Weir, *Ent. Rec.* **92**:218).

1982 – **Ranmore Common,** 9.5 (MSH).

Nymphalis antiopa (Linnaeus, 1758) PLATE 12 **Camberwell Beauty**

Migrant; rare.

The Camberwell Beauty is a migrant species which comes to us from Scandinavia and northern Europe rather than having a Mediterranean origin as have most other migrant butterflies. Consequently, as migration is to no small degree controlled by climatic conditions, the Camberwell Beauty's appearance is more irregular and much less frequent than other species, but occasionally it appears in large numbers. The largest migration was in 1872 in which year thirteen examples were noted in Surrey (Barrett, 1893), and the most recent large scale invasion was in 1976. Evidence suggests that some of the 1976 migrants were able to survive the winter to appear in the spring of 1977, although there has never been any evidence of breeding in this country. In the late 1950s L. Hugh Newman made several fairly large scale releases, of marked butterflies, in areas not too far distant from Surrey: 1956, Greenwich Park, West Kent (50 adults); 1958, Letchworth, Herts. (150 adults); 1959, Lullingstone, West Kent; 1960, Westerham, West Kent (60 adults). Despite three of these sites being within a few miles of the Surrey border only one individual was noted in the county during those years. The vernacular name is derived from an early record of two examples at Camberwell, Surrey in August 1748.

1858 – **Weybridge**, 6-7.10 (2) (Pennell, *E.W.I.* **5**:26).

1872 – 13 in Surrey (Barrett, 1893).

1875 – **St. Ann's Hill**, 10.8 (Wailly, *Ent.* **8**:197).

1881 – one in April, [locality not given] (South, 1906).

1888 – one, [locality not given] (South, 1906).

1889 – one in April, [locality not given] (South, 1906).

1891 – **Balham**, September (South, 1906).

1896 – **Epsom**, December (South, 1906).

1900 – **Oxshott**, 20.4 (Hewat, *VCH*); **Woking**, August (Saunders, *E.M.M.* **37**:100).

1904 – **Raynes Park**, 29.8 by W. Smith (*Ent.* **38**:91).

1910 – **Weybridge**, 8.3 (Parker, *Ent.* **43**:119).

1915 – **Addiscombe**, 22.9 (Saville, *Ent.* **48**:267).

1917 – **Oxshott**, 2.8 (Dallas, *Ent.* **50**:235).

1924 – **Reigate**, 25.9 by H. Speyer (*Ent.* **57**:282).

1928 – **Dulwich Park**, 2.9 (Frohawk, *Ent.* **61**:236).

1929 – **Virginia Water**, 13.8 (*Ent.* **63**:14).

1930 – **Egham**, 8.3 (*Ent.* **63**:163).

1934 – **Ashtead**, 24.5 by A.W. Buckstone (*Ent.* **68**:15).

1943 – **Sheerwater**, 4.7 (*Ent.* **76**:154).

1944 – **Guildford**, 18.6 (Parfitt, *Ent.* **78**:26).

1945 – **Farnham**, 26.8 (*Ent.* **78**:144).

1946 – **Ashtead Woods**, spring (Russwurm); **Reigate** golf course, 7.4 (Seth-Smith, *Ent.* **79**:162).

1947 – **Kenley** (*Ent.* **80**:268); and another in Surrey (Migration records, *Ent.* **81**:112).

1949 – **Farthing Downs**, one (Bigger, *Ent. Gaz.* **10**:133).

1953 – **Camberley**, 1.7 (*Ent.* **87**:62).

1959 – **Chobham**, 30.8, possibly originating from those released by L.H. Newman at
Lullingstone, Kent, 24.7.1959 *(Ent.* **95**:175).

1963 – **Dorking**, 3.10 *(Ent. Rec.* **76**:119).

1968 – **Wentworth**, 7.9 (Howarth, *Proc. BENHS* **1**:112).

1976 – **Godalming**, July; [**Ewshot**, 22.8, in Hants not Surrey]; **Onslow Village,
Guildford**, 22.8; **Richmond**, 23.8; **Kew**, 24.8; **Dulwich**, 27.8;
Banstead Heath, late August; **Dulwich**, 3.9; **Carshalton**, 3.9;
Godalming, 4-5.9 (Chalmers-Hunt, *Ent. Rec.* **89**:101); **Witley**, 7.9;
St. Martha's Hill, 18.9 (Chalmers-Hunt, *Ent. Rec.* **89**:249).

1977 – **Netley Heath**, 27.4 (Chalmers-Hunt, *Ent. Rec.* **90**:87); **Leatherhead**, 22.5
(Chalmers-Hunt, *Ent. Rec.* **89**:248) – probably both survivors from the previous year.

1981 – **South Croydon**, 15.8 (Smith, *Ent. Rec.* **93**:241); **Kew Gardens**, 25.8 (R. Hastings).

1994 – **Bletchingley**, 3.8 (Mrs S. Ruck per R. Wason).

Inachis io (Linnaeus, 1758)   PLATES 12,9   Peacock

Resident; commons, gardens,
woodland rides; wide-
spread and common.

Univoltine; late July to early
September, and after
hibernation in April
and May.

Foodplant – common nettle.

The Peacock is a species that occurs
widely in Surrey in a variety of
habitats and is frequently seen in
domestic gardens. Over the last few
years numbers have been very high,
but prior to that there was a period

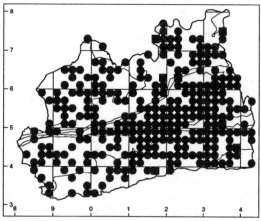

of relative scarcity. Such fluctuations are a feature of most butterfly populations, being
connected with climatic conditions and interactions with parasites, and are not a cause for
concern – as long as suitable habitat exists together with terrain over which recolonization
may occur, numbers are able to build up as quickly as they have declined. Like the other
native members of the Nymphalinae, the Small Tortoiseshell and the Comma, the Peacock
hibernates as an adult and is regularly seen in the wartime pill-boxes and forts of the North
Downs. In an old Napoleonic fort on Box Hill, where the hibernation of *Triphosia dubitata*
(Tissue Moth) was being studied, it was noted that whereas the moths were hibernating in
the subterranean level where the temperature remained constant throughout the winter, the
butterflies (Peacock and Small Tortoiseshell) utilized the ground floor and had gone by
late winter (Morris and Collins, 1991). The adult is, except in extremely rare circumstances,
single-brooded and will hibernate fairly soon after emergence, only rarely being seen after
the end of August. Indeed individuals have been found roosting as early as July, but it is
not known whether they would have reappeared later or would have remained torpid until
the spring.

Pictures of adults follow the checklist order.
Plates 8 and 9 show larvae and ova.

Thymelicus sylvestris   Small Skipper

Thymelicus lineola   Essex Skipper

Hesperia comma   Silver-spotted Skipper

Ochlodes venata   Large Skipper

PLATE 1

SHIRLEY WHITE

Erynnis tages   Dingy Skipper

TONY HOARE

Pyrgus malvae   Grizzled Skipper

TONY HOARE

Papilio machaon   Swallowtail

GAIL JEFFCOATE

Leptidea sinapis   Wood White

PLATE 2

Colias croceus   Clouded Yellow

Gonepteryx rhamni   Brimstone

Pieris brassicae   Large White

Pieris rapae   Small White

PLATE 3

Pieris napi Green-veined White

Anthocharis cardamines Orange-tip

Callophrys rubi Green Hairstreak

Thecla betulae Brown Hairstreak

PLATE 4

Quercusia quercus   Purple Hairstreak

Satyrium w-album   White-letter Hairstreak

Satyrium pruni   Black Hairstreak

Lycaena phlaeas   Small Copper

PLATE 5

Lampides boeticus Long-tailed Blue

Cupido minimus Small Blue

Plebejus argus Silver-studded Blue

Plebejus argus Silver-studded Blue

PLATE 6

KEN WILLMOTT

Aricia agestis   Brown Argus

PAUL UNDERWOOD

Polyommatus icarus   Common Blue

MIKE WELLER

Lysandra coridon   Chalk-hill Blue

BARRY HILLING

Lysandra bellargus   Adonis Blue

PLATE 7

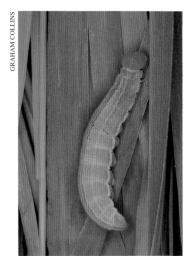

Family Hesperiidae   Small Skipper

Family Papilionidae   Swallowtail

Family Pieridae   Brimstone

Family Lycaenidae   Brown Hairstreak

PLATE 8

Family Lycaenidae   White-letter Hairstreak

Family Lycaenidae   Holly Blue

Family Lycaenidae   Duke of Burgundy

Family Nymphalidae   Peacock

Family Nymphalidae   Meadow Brown

PLATE 9

Celastrina argiolus   Holly Blue

Hamearis lucina   Duke of Burgundy

Ladoga camilla   White Admiral

Ladoga camilla   White Admiral

PLATE 10

Apatura iris   Purple Emperor

Vanessa atalanta   Red Admiral

Cynthia cardui   Painted Lady

Aglais urticae   Small Tortoiseshell

PLATE 11

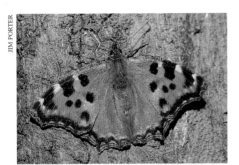

Nymphalis polychloros   Large Tortoiseshell

Nymphalis antiopa   Camberwell Beauty

Inachis io   Peacock

Polygonia c-album   Comma

PLATE 12

Boloria selene   Small Pearl-bordered Fritillary

Boloria euphrosyne   Pearl-bordered Fritillary

Argynnis lathonia   Queen of Spain Fritillary

Argynnis adippe   ◄ High Brown Fritillary

PLATE 13

PAUL UNDERWOOD

Argynnis aglaja   Dark Green Fritillary

DAVID DUNBAR

Argynnis paphia   Silver-washed Fritillary

GRAHAM COLLINS

Eurodryas aurinia   Marsh Fritillary

TONY HOARE

Pararge aegeria   Speckled Wood

PLATE 14

Lasiommata megera   Wall

Melanargia galathea   Marbled White

Hipparchia semele   Grayling

Pyronia tithonus   Gatekeeper

PLATE 15

GAIL JEFFCOATE

Maniola jurtina   Meadow Brown

GAIL JEFFCOATE

Aphantopus hyperantus   Ringlet

JIM PORTER

Coenonynpha pamphilus   Small Heath

TONY HOARE

Danaus plexippus   Monarch

PLATE 16

Polygonia c-album (Linnaeus, 1758) PLATE 12 Comma

Resident; commons, gardens, woodland rides; widespread and fairly common.

Partially bivoltine; peaks in April and July.

Foodplant – common nettle, hop, elm, currant, gooseberry.

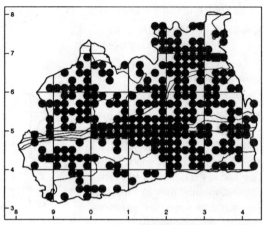

The Comma has undergone the most remarkable decline and later expansion in range of any British butterfly. At its nadir in about 1913 it could be found only on the English/Welsh border and was entirely absent from Surrey, but by 1930 had been reported from Bookham and Cheam (Turner, *Ent. Rec.* **42**:155), and by the 1950s was described as "recently recorded from all over the London area" (de Worms, 1950), "well distributed and fairly common in north-west Surrey" (Bretherton, 1957), and "never a common species, but turns up here and there" (Perrins, 1959). Its current status is of a widespread species found throughout the county and occurring well into London. Numbers are rarely high and the early stages are seldom seen, the larvae being solitary, although no doubt the fairly catholic choice of foodplants allows the butterfly to breed in a wide variety of habitats. The adult hibernates, reportedly in the open on branches and tree trunks, and is one of the earlier butterflies to be seen on the wing in the spring. The only record of hibernation was an example found resting, head down and about five feet from the ground, on the south-easterly side of a trunk of an oak tree at Ashtead Common, 4.2.95 (GAC). From the eggs laid in the spring some larvae feed up rapidly to produce butterflies of the form *hutchinsoni* which produce a second brood in the autumn, while some feed more slowly producing adults in the late summer which will hibernate along with the second brood.

Araschnia levana (Linnaeus, 1758) Map Butterfly

Escape from captivity or vagrant.

A single example of this species was swept from bilberry undergrowth at Friday Street, 21.5.1982, and included by Bretherton & Chalmers-Hunt in their migration records for that year. They considered it as probably a migrant as a considerable migration of Red Admiral and other species was in progress at the time. However, apart from a known introduction in 1912, it has not been previously recorded from Britain and, although spreading northwards in Europe, is not generally considered as a migrant species. Additionally it is a species commonly available from dealers in livestock, of which there are several in the vicinity of the capture, and an escape from captivity would seem a far more probable reason for its occurrence in Surrey. It should be pointed out that the deliberate release of specimens is forbidden by the Wildlife and Countryside Act.

Boloria selene ([Denis and Schiffermüller], 1775) **Small Pearl-bordered Fritillary**

ssp. *selene* ([Denis and Schiffermüller], 1775)   PLATE 13

Resident; open woodland, damp heathland; very local and uncommon.

Univoltine; June.

Foodplant – ([marsh violet]).

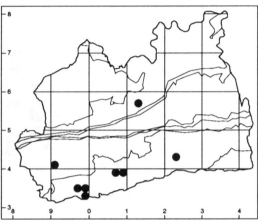

The Small Pearl-bordered and, to an even greater extent, the Pearl-bordered Fritillary have suffered a severe decline in the south-east of England in the last 30 to 40 years, a decline mainly due to changing woodland management practices. Both species are currently restricted to the extreme south-west of the county, principally to the woodland around Chiddingfold, where the populations remain fairly stable. In addition the Small Pearl-bordered Fritillary has been introduced to a wood in northern Surrey where more traditional woodland management practices have been re-established, and appears to be surviving well. Occasional recent records for the London area are considered to be due to release of breeding stock. The butterfly was formerly much more widespread, although commoner in the damper areas of the west – de Worms (1950) considered *selene* to be much less widespread in the London area than *euphrosyne* listing only Hinchley Wood and Prince's Coverts; Evans (1973) agreed adding only Ashtead Woods. Referring to colonies in north-west Surrey, Bretherton (1957) wrote "the usual foodplant here appears to be *Viola palustris*" – whether this was based on inference or direct observation is not known; common dog-violet is much more likely to be utilized in its current colonies. The two species are superficially very similar and considerable care should be taken when recording them.

RECORDS – **Thursley Common**, 21.6.80, 21.6.86 (PFC), 1992 (R. Fry);
Chiddingfold, 15.6.85 (PFC), 17.6.85 (GAC), 19.6.88 (RKAM);
Tugley Wood, 20.6.82 (TJD), 8.6.84 (BC), 31.5-2.6.85, 10.6.87 (SCP),
18.6.94 (P. Farrant); **Botany Bay**, 15.6.85 (PFC), 22.6.94 (M. Bennett);
Bowles Wood, 1992 (DWB); **Lower Canfold Wood**, 10.6.92 (DWB);
Bookham Common, 6.7.85 (D. A. Boyd (per CWP)); **Edolphs Copse**, 3.6.87 (GAC).

Boloria euphrosyne (Linnaeus, 1758)   PLATE 13 **Pearl-bordered Fritillary**

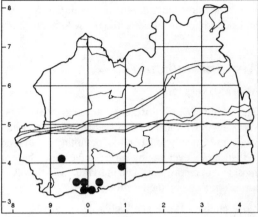

Resident; open woodland; very local and scarce.

Univoltine; mid-May to June.

In Surrey, as in much of south-east England, the Pearl-bordered Fritillary has declined from being a common and widespread woodland species to virtual extinction in a period of less than 40 years. Its present distribution follows closely that of the Small Pearl-bordered Fritillary (q.v.). This species shows a preference for drier habitat than the previous one and was once commoner in the east of the county. The larva requires violet growing in open, sunny situations where it can bask in spring sunshine; this severe decline is almost certainly due to the cessation of coppicing and subsequent shading. In the middle of the century de Worms (1950) describes it as often abundant in woods of the London area, giving for Surrey: common around Epsom, Ashtead Woods, abundant in Oxshott woods, common on Surbiton Golf Course, and a few at Caterham; also (de Worms, 1959): numerous on Bookham Common, Chipstead and Mickleham Downs, Wimbledon Common, one in May 1945. For north-east Surrey Evans (1973) concurred, although he had no records later than 1958, and in the north-west Bretherton (1957) considered it local but not common. Recently it has occurred at two additional sites not shown on the map, almost certainly as a result of undocumented introduction as they are a considerable distance from known colonies. Second brood examples are exceedingly rare, being recorded as follows: Haslemere, 15.7.1868 (Barrett, 1893); Dunsfold, 8.9.1989 (PJH).

RECORDS – **Witley Common**, 1980 (2) (DWB); **Chiddingfold**, 15.6.85 (PFC); **Tugley Wood**, 20.6.82 (TJD), 8.6.84 (BC), 16.5-2.6.85, 2-19.6.86 (SCP), 8.6.92 (GAC), 14.6.94 (G. Revill); **Botany Bay**, 30.5.84 (PFC), 17.6.85 (GAC), 11.5.93 (G. Revill), 11.6.94 (M. Bennett); **Dunsfold**, 8.9.89 (PJH); **Alfold**, 20.5.92 (DWB); **Sidney Wood**, 20.6.85 (IDMacF); **Canfold Wood**, 28.5.92 (MR).

Boloria dia (Linnaeus, 1767)   **Weaver's Fritillary**

Introduction.

A male specimen of Weaver's Fritillary was taken at Westcott Downs on August 24th 1984 and later exhibited at the annual entomological exhibitions of that year. The specimen had resulted from the release of some fifty full grown larvae of continental origin at the site about a month earlier. An earlier example is of a female taken at Worcester Park by W.A. Smith in 1872 (Lewis, *E.M.M.* **12**:229), one of four or five considered by Barrett (1893) to be acceptable records and which he suggested had been accidentally introduced with imported plants. Goss however, in the *VCH*, considered it more likely to have resulted from confusion with a continental example. Weaver's Fritillary is not native to this country and should not be released into the wild in any circumstances.

Argynnis lathonia (Linnaeus, 1758)   PLATE 13   Queen of Spain Fritillary

Migrant; very rare.

The Queen of Spain Fritillary is an occasional migrant which is resident in parts of western Europe. As with many other scarcer migrants it has been a less frequent visitor in recent years, presumably as a result of declining populations in its breeding grounds. There is no evidence of it ever having survived the winter here and it is unlikely to establish itself. The most recent record was of an individual which was seen and photographed over a period of a week at a site on the North Downs.

19th century – **Battersea Fields**, 1818 (Stephens, 1828). **Headley, Betchworth** and **Redhill**, prior to 1851 (*VCH*). **Croydon**, second week of August, 1868 (Newman, 1874).
1934 – **Warlingham**, 30.9 a female (Owen, *Ent.* **67**:248).
1943 – **Sutton**, 17.8 (Frohawk, *Ent.* **78**:21).
1949 – near **Haslemere**, 2.9 (Tidmarsh, *Ent.* **82**:250).
1964 – **Ockley Common**, 25.7 (*Ent.* **99**:237).
1976 – "**North Downs**", 12-19.7 (Willmott, *Ent. Rec.* **88**:333).

Argynnis adippe ([Denis and Schiffermüller], 1775)   High Brown Fritillary
ssp. *vulgoadippe* (Verity, 1929)   PLATE 13

Extinct.

Barrett (1893) considered that the High Brown Fritillary occurred in most of the larger woods of the southern counties but added that at Haslemere, a hilly and rather cool neighbourhood, it was exceedingly scarce. In north-west Surrey it was also very scarce with only six examples recorded from 1934 to 1948 (Bretherton, 1957), a period during which it had become abundant elsewhere. De Worms (1950) listed Ashtead Woods, Prince's Coverts and Wimbledon Common, 1947. Perrins (1959) described it as frequent in the Chiddingfold area, but scarcer now; he had only recorded two examples: 7.7.1953 and July 1955. Central Surrey was the stronghold of this species from the time of the *VCH* until about 1960, for example: Norbury Park, 8.7.1951 (ASW); Ashtead, 25.6.52, 19.6.59 (SFI); Bookham Common, most years 1951-59 (ASW). After this period it declined rapidly, not just in Surrey but throughout the south-east of England and is presently to be found only in western and north-western England. This decline is thought to be due to changing woodland management practices, especially the cessation of coppicing, although where it still occurs it feeds on violet growing beneath quite dense bracken stands. Because of this current scarcity it is afforded protection on schedule five of the Wildlife and Countryside Act (1981).

Argynnis aglaja (Linnaeus, 1758)   PLATE 14   **Dark Green Fritillary**
ssp. *aglaja* (Linnaeus, 1758)

Resident; chalk grassland;
          restricted and
          increasingly uncommon.
Univoltine; late June to mid-
          August.

The Dark Green Fritillary is
essentially a species of chalk
grassland, at least in the south-east
of England, although in years of
plenty it can spread and temporarily
colonize open deciduous woodland.
Wheeler (1955) gave it as occurring
on Bookham Common, but less
frequent than the High Brown, and

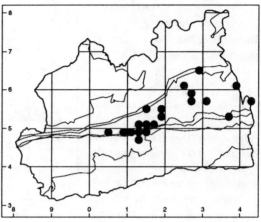

it was noted, not uncommonly, in the Chiddingfold Woods on the Sussex side of the border
in the late 1970s (S. Church, pers. com.). The example noted at Beddington Sewage Farm
on 19.7.1987 (DC) probably arrived there by natural means; it is known to occur at Banstead
Downs some four miles distant. Numbers do tend to fluctuate and by the early years of the
1990s it had become quite scarce, although in the last two years it seems to have been
staging a comeback. In common with most of the Fritillaries there was a considerable
decline after about 1950 but by the early 70s it had increased again and at some sites was
as common as it had ever been (Evans, 1973). Much of the change in frequency must be
due to climatic factors, although the availability of suitable foodplant, usually common
dog-violet and hairy violet, may also be important; lack of grazing by rabbits after the
myxomatosis epidemic may have resulted in the larval foodplant being too sheltered;
conversely overgrazing by rabbits, now at very high population density again, could lead
to loss of violet plants. In his work on British lepidoptera Barrett wrote "in Surrey I found
it very scarce" (Barrett, 1893) – possibly his experience of the butterfly was gained in the
woods around Haslemere where he lived as the *Victoria County History* described it as
generally distributed on the chalk downs in the centre of the county.

RECORDS – **Newlands Corner**, 9.7.86 (ASW); **Hackhurst Downs**, 8.7.81 (PFC);
White Downs, 11.7.82 (TJD), 1.7.87 (RDH), 28.7.90 (T. Smithers); **Wotton**, 1982 (JDH);
Westcott Downs, 22.8.90, 15.8.94 (GJ); **Ranmore**, 29.6.80, 26.6.82 (TJD), 21.7.86 (PFC),
18.7.94, 22.8.90 (GJ); **Dorking**, 19.6.84, 1.8.86 (PC); **Norbury Park**, 1984-86 (SBRC);
Box Hill, 1994 (A. Reid); **Juniper Top**, 27.6.92, 29.6.93, 10.7.94 (P. Churchill);
Headley Warren, 25.7.92, 27.6.93 (HM-P); **Banstead Downs**, 1.8.85 (AAW),
1990, 94 (DC); **Park Downs**, 24.7.94 (S. Price); **Chipstead Valley**, 16.7.84 (GAC);
Beddington, 19.7.87 (DC); **Farthing Downs**, 13.7.85, 29.7.86 (ASW), 31.7.87 (PFC),
24.7.91 (N. Brown); **Happy Valley**, 21.7.84 (PFC); **Addington**, 5.7.88, 29.6.90,
25.6.93 (PFC), 11.7.93 (GJ), 7.8.94 (A. Shelley); **South Hawke**, 6.8.84 (GAC);
Tatsfield, 13.7.87 (GAC).

Argynnis paphia (Linnaeus, 1758)   PLATE 14   **Silver-washed Fritillary**

Resident; open woodland;
    restricted but fairly
    common.

Univoltine; late June to August.

Foodplant – (common dog-
    violet).

The Silver-washed is our largest
species of Fritillary and a
magnificent sight gliding along the
rides of its woodland habitats.
Nationally it is contracting its range
in the east, but in Surrey the opposite
would appear to be the case; in
addition to most of the wealden

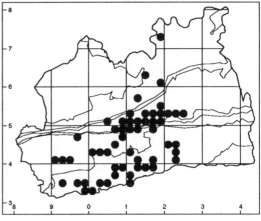

woods where it is fairly common, it can also be found in many of the woods and even open
downland on the chalk and has recently been found in a number of sites from where there
have been few records since the 1950s. The *VCH* gave it as being widespread but as occurring
only sparingly, not being as abundant in Surrey as it was in the New Forest; it is certainly
the reverse today. The green form of the female, f. *valesina* Esper, which forms a significant
proportion of the population in some Hampshire and Wiltshire colonies, occurs in Surrey
only as a rare aberration and has been noted as follows: Tugley Wood, 11.7.1982 (PJH)
and Chiddingfold, 18.7.1982 (MSH) [possibly the same specimen]; Glovers Wood, 4.8.1990
(GAC). Its occurrence in Surrey is evidently not a new phenomenon as several examples
were recorded in the Cranleigh area from 1948 to 1951 (*teste* Perrins, 1959). A bilateral
gynandromorph was recorded from a west Surrey wood, 4.8.1976 (RFB).

Eurodryas aurinia (Rottemburg, 1775)   PLATE 14   **Marsh Fritillary**

Extinct resident and introduction.

The history of the Marsh Fritillary in Surrey is confused by numerous attempts to introduce
livestock and by escapees from captivity. It was certainly once resident, possibly until as
recently as 1973, but has now been lost as a Surrey species. Nationally there has been a
westwards decline during much of this century, related to the massive loss of damp grassland
habitats. The *VCH* gives it from Haslemere, adding that it had not been reported recently
and had probably disappeared from Surrey. In north-west Surrey there were colonies at
Fairoaks, Lucas Green and Clasford Common as well as at Littlefield Common (Bretherton,
1957). In the south-west it was known from Little Frensham Pond and Hankley golf course,
June 1951 (Perrins 1959). The butterfly survived for a few years on Bagmoor Common,
having been introduced in 1965 (DWB). Probably the last area where the Marsh Fritillary
occurred naturally in Surrey was at Littlefield Common. Over 20 adults were seen there on
5.6.1970, and one on 28.5.1971 (DWB). When C.J. Luckens visited the site in 1972 he
saw no butterflies and discovered that the common had been badly burnt leading him to

fear that the colony had been destroyed. However, in August 1973 the site was again visited and a larval nest found from which some twenty larvae were taken for captive breeding. In 1974 fertile females resulting from these larvae were taken to the site but, as no wild butterflies were seen and the scabious [devil's-bit] was severely depleted, re-introduction was not attempted. In 1976 larvae were introduced to Lower Canfold Wood and in 1978 to Fisherlane Wood (Luckens/RFB) but at neither site were they seen again. A further introduction was attempted at Chiddingfold where they were seen for a few years from 25.5.1982 (MSH) until 30.4.1986 when a solitary larva was found (GAC). Lack of success at these sites was probably related to their being insufficiently open as the larvae require ample sunshine for their post-hibernation development. Conservation of this species would be better served by protecting the habitat in which it still occurs naturally than by indiscriminate, and frequently undocumented, releases in areas in which it is most unlikely to survive.

Melitaea cinxia (Linnaeus, 1758) Glanville Fritillary

Extinct.

At one time the Glanville Fritillary was rather more widespread than at present and amongst the localities at which it occurred was Dulwich, leading Petiver, who had originally called it the Lincolnshire Fritillary, to rename it in 1717 as the Dullidge Fritillary. It was probably lost from Surrey in the eighteenth century and by the middle of the nineteenth century had retreated to the south coast of England, where it is now restricted to the south coast of the Isle of Wight.

Pararge aegeria (Linnaeus, 1758)   PLATE 14   Speckled Wood
ssp. *tircis* (Godart, 1821)

Resident; woodland, commons, gardens; widespread and common.

Bi- or trivoltine; several overlapping broods, April to October.

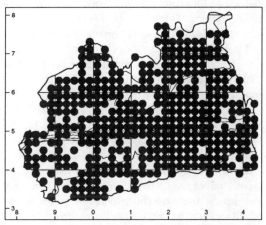

Like the Comma, the Speckled Wood has undergone a very great contraction and a later expansion in range over the last hundred years. Newman (1874) considered it an abundant species in almost every part of the country. In the late nineteenth and early twentieth centuries the Speckled Wood declined seriously becoming confined to a few areas in southern and south-western England and Wales. MBGBI 7(1) considered it to have become confined in the south-east to West Sussex, but it probably occurred in Surrey throughout that period, being described as "common in wooded parts of Surrey" (Barrett, 1893), although in many other places Barrett considered it to have declined severely, "generally common in woods and lanes" (*VCH*), and not uncommon in the Witley district 1910s (Tullett, *Ent. Rec.* **32**:52). Currently it is the most widespread species in the county being found not uncommonly in open areas as well as woods, and even breeding in suburban gardens, although as it is a territorial species it does not occur in large numbers. The winter can be passed as either a larva or as a pupa and the butterfly occurs almost continuously from April to October in a number of overlapping broods.

Lasiommata megera (Linnaeus, 1767)   PLATE 15   **Wall**

Resident; commons, lanes, woodland rides; restricted and uncommon.

Bivoltine; mid-May to June, and mid-July to August.

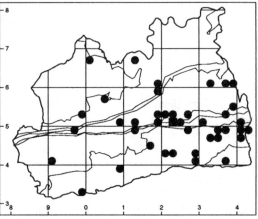

The history of the Wall butterfly is the complete opposite of that of the previous species and it has of late seriously declined within the county. At the turn of the century it was described as "common every-where by roadsides throughout the county" (*VCH*). In the 1940s and 50s de Worms (1950) gave many sites including "common at Oxshott, widespread in the Box Hill district, and numerous in Richmond Park". It was also common at Bookham (Wheeler, 1955), although Bretherton (1957) considered it only fairly common in the north-west, while Evans (1973) gave it as generally distributed and often common in north-east Surrey, and it would seem that the decline is recent. During the survey period it has, with one or two exceptions, been confined to the wealden lanes of the south-east, from where there are a number of records for the first half of the 1980s, and to the vicinity of the chalk, although records for the past few years suggest that it may be increasing again. Records for the current decade being: Westcott Downs, 22.8.90 (GJ); Epsom Common, 28.6.92 (PFC); Sheepleas, 1993 (D. Smith); Field Common, 18.8.93 (A. Petrie); Ewhurst, 1994 (SFI). A probable factor in its decline is climatic, the butterfly having experienced similar declines inland in a number of neighbouring counties while remaining not infrequent in coastal localities. In Hampshire its decline is attributed to a series of cold wet summers in the second half of the 1980s (Goater, 1992) and the pattern of decline in Surrey fits well with this theory. There are still populations to the north-west of the county, and in Sussex it would appear to be spreading again, so there is every probability that the Wall will recolonize the county in the next few years, climatic conditions permitting.

Melanargia galathea (Linnaeus, 1758)   PLATE 15   **Marbled White**
ssp. *serena* Verity, 1913

Extinct resident and introduction; local but common.
Univoltine; July to mid-August.

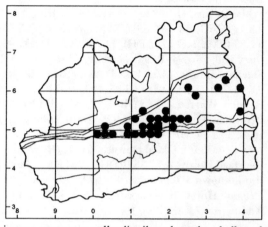

The exact history of the Marbled White in Surrey is difficult to determine, being confused by both numerous, fairly recent, intro-ductions and conflicting records of past frequency. The apparent picture is of a species formerly resident but only very locally common, which possibly became extinct after 1950, and has more recently occurred only as a result of various introductions and natural spread from these, but is now more generally distributed on the chalk and frequently very common. A specimen from Surrey in 1870 was amongst those exhibited by Platt-Barrett at a meeting of the SLENHS (*Proc. SLENHS* **1915-16**). C.G. Barrett (1893) described it as "more local and scarce in...Surrey", adding that, at Haslemere, he had only seen single examples on two or three occasions. For the same period the *VCH* gives it as occurring only sparingly at: near Godalming, Mickleham, Gomshall and Ranmore Common, and Sheepleas. One was seen at Betchworth, 1946 (RF), and de Worms (1950) gives one at Cobham, 1947, and one at Hinchley Wood, 1949 [these sites are within a few miles of each other]. Perrins (1959) recorded one flying past Charterhouse in July 1949, adding that it occurred on Box Hill in numbers. Writing in 1959 Collier (*Ent. Rec.* **71**:119) stated that he had never been able to understand why it did not appear to exist in Surrey, adding "I hope that nobody's feelings will be hurt if I introduce them again into parts of the North Downs". According to Oates and Warren (1990) it was released on Ranmore Common [i.e. Westcott Downs] in July 1964 (but not by Collier), as females originating from Kent, existing within a small area for a few years before becoming more widespread in the early 1970s. In 1969 two were seen at Juniper Bottom and a further one at Westcott Downs (RF), the only ones he had seen in Surrey apart from the one at Betchworth. Further examples were released at Headley in the 1970s and it is very common there today. The centre of distribution is on the scarp slope of North Downs between Guildford and Reigate, although the most extreme westerly colonies have only established in the last few years, and a number of smaller colonies occur in areas on the dip slope of the chalk. These latter colonies are the result of various known introductions, some successful, others dying out after only a few years. Odd specimens have also been recorded from the Surrey/Hampshire border. A colony in the area of Windsor Forest, Berkshire, probably introduced (Oates and Warren, 1990) was likely to be the source of a record from Chobham Common, 26.7.75 (PAM) .

RECORDS – **Bookham Common**, 20.7.85 (D. A. Boyd (per CWP));
Pewley Down, 15.7.94 (M. Bennett); **Merrow Downs**, 1992 (E. Wood);

Newlands Corner, 8.7.90 (E. Wood), 15.7.94 (M. Bennett); **Sheepleas**, 1994 (D. Smith);
East Horsley, 29.7.94 (GAC); **Colekitchen Down**, 3.8.94 (GJ);
Hackhurst Downs, 2.8.81 (RDH), 9.8.84, 9.8.91 (GAC), 3.8.94 (GJ);
White Downs, 11.7.82 (TJD), 31.7.94 (B. Hilling); **Westcott Downs**, 13.7.83, 9.8.84,
9.8.91 (GAC), 14.7.94 (GJ); **Ranmore**, 29.6.80, 26.6.82 (TJD), 9.8.84, 17.7.85 (GAC),
6.8.85 (EM), 11.7.87 (RDH), 14.7.94 (GJ); **Dorking**, 11.7.87 (RDH), 21.7.94 (GJ);
Bury Hill, 25.7.94 (RDH); **Norbury Park**, 13.7.85 (PFC);
Fetcham Downs, 14.7.94 (D. Dunkin); **Box Hill**, 18.7.82 (TJD), 14.5.83 larvae,
9.8.84 (GAC), 23.7.84, 13.7.85, 19.8.91 (PFC), 23.7.94 (ME);
Headley Warren, 7.7.85 (JBS), 27.6.93 (HM-P), 9.7.94 (GAC);
Brockham Quarry, 14.8.94 (GJ); **Dawcombe**, 25.7.92 (RDH), 17.7.94 (GJ); **Buckland
Hills**, 30.7.94 (GJ); **Colley Hill**, 26.7.86 (Mrs P. Dawe (per CWP)),
4.7.94 (A. Reid); **Oxted Downs**, 26.6.92 (H. Whiting);
Banstead Downs, 24.7.88 (RDH), 1994 (DC); **Park Downs**, 29.6.93 (A. Reid);
Riddlesdown, [20 introduced: 18.7.82], 18.7.84 (PFC), 8.7.85 (GAC);
Croham Hurst, [introduced: 20.7.78], 30.7.80, 12.7.81 (PFC);
Addington, 7.8.83 (BC), 16.7.85 (GAC), 27.7.91, 18.7.92, 25.6.93, 10.7.94 (PFC).

Hipparchia semele (Linnaeus, 1758)   PLATE 15   Grayling
ssp. *semele* (Linnaeus, 1758)

Resident; heathland; local but fairly common.

Univoltine; mid-July to early September.

The Grayling is a resident of the larger heaths in the west of the county, and still occurs in good numbers. It does though appear to be absent from the more outlying areas of heathland such as at Wisley Common and Blackheath. The adult is usually to be found resting on bare sandy ground where its underside coloration and habit of leaning at an

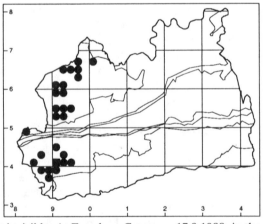

angle to the ground render it almost invisible. At Frensham Common, 17.8.1988, in the late afternoon, adults were observed resting on the trunks of pine. This may have been due to the low angle of the sun, but it is equally likely that they roost in trees at night. In addition to its heathland localities, the Grayling was a former resident of the chalk downs but became extinct there by the early 1960s. Amongst some of the later records were Westcott Downs, 19.7.1959 (*Proc. SLENHS* **1959**) and on the downs near Dorking, 1962 (Collier, *Ent. Rec.* **75**:33). The reason for this local extinction is not clear, especially as numbers have held up well on the heaths, but may have been due to scrub invasion following myxomatosis depriving the adult of bare ground on which to settle; it probably also breeds around the edges of bare areas. It would also appear to have occurred occasionally on Wimbledon and Bookham Commons in the 1950s (de Worms, 1950).

RECORDS – **Hale Common**, 21.9.80 (RDH); **Hindhead**, 20.8.94 (DWB); **Frensham Common**, 7.8.87 (RKAM), 17.8.88 (GAC), 16.7.94 (M. Bennett); **Devils Jumps**, 5.8.89, 11.8.90 (GAC); **Tilford**, 31.8.86 (GAC); **Hankley Common**, 20.8.94 (DWB); **Thursley Common**, 13.5.81, larvae (SHC), 1.8.84 (JBS), 8.8.84 (GAC), 20.7.88 (PFC), 15.8.94 (D. Dell); **Bagmoor Common**, 15.8.92 (B. Whitty); **Witley Common**, 1.8.85 (IDMacF); **Puttenham Common**, 6.8.88 (RKAM); **Wyke Common**, 14.8.94 (GAC); **Tunnel Hill**, 3.8.94 (J. Pontin); **Cleygate Common**, 29.7.89 (J. Pontin); **Henleypark Lake**, 7.8.94 (D. Dell); **Frith Hill**, 15.8.87 (GAC); **Bisley Ranges**, 27.7.92 (J. Pontin); **White Hill**, 7.8.88 (GAC); **Pirbright Common**, 7.7.83 (PC); **Pirbright**, 14.8.94 (GAC); **Brentmoor Heath**, 21.7.80 (PJH), 15.7.87 (PFC), 1990-94 (M. Adler); **Lightwater**, 3.9.91 (DWB); **Windlesham**, 13.8.94 (GAC); **Chobham Common**, 24.7.85 (PFC), 9.9.87 (RDH), 13.7.85, 1.8.87, 19.9.91 (GAC), 31.7.94 (B. Gerrard); **Thorpe**, 31.7.82 (PJB).

Pyronia tithonus (Linnaeus, 1771) PLATE 15 Gatekeeper
ssp. *britanniae* (Verity, 1914)

Resident; grassland, woodland rides, heathland; widespread and common.

Univoltine; mid-July to August.

Foodplant – grasses.

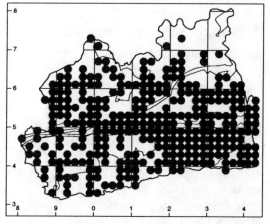

The Gatekeeper is a widespread and common species in Surrey, being found in most areas of grassland, whether on the chalk, the weald or amongst heathland, and absent only from the more built-up parts of London where it is nevertheless frequent at such sites as Richmond Park and Mitcham Common. It is particularly fond of flying along the line of a hedgerow and will congregate in numbers about any gaps giving rise to the preferred English name and its almost equally well known alternative, the Hedge Brown. In common with several other species of Brown, the larvae can be readily swept from grasses after dark in the spring.

Maniola jurtina (Linnaeus, 1758)   ssp. *insularis* Thomson, 1969

PLATES 16,9

Meadow Brown

Resident; grassland, woodland rides; widespread and common.

Univoltine; mid-June to August.

Foodplant – grasses.

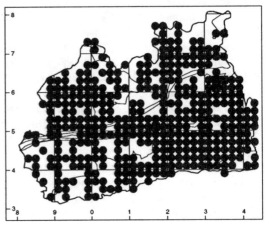

The Meadow Brown is one of the most widely distributed butterflies in Surrey, only the Speckled Wood being recorded from a larger number of tetrads. It is also probably one of the most abundant species, occurring in large numbers on almost any patch of grassland, including such areas that are left undisturbed in domestic gardens. The flight period of the adult is rather prolonged, this being due to extended periods of egg laying and the ability of the larvae to feed during suitably warm periods during the winter; the adult is strictly univoltine. The larvae are easy to sweep in numbers after dark in the late spring.

Aphantopus hyperantus (Linnaeus, 1758)

PLATE 16

Ringlet

Resident; grassland, woodland rides; fairly wide-spread and common.

Univoltine; July and early August.

Foodplant – grasses.

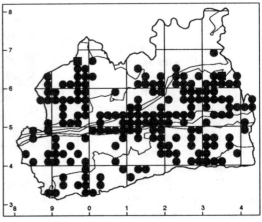

The Ringlet is fairly widely distributed in Surrey, although it is virtually absent from the London area, eschewing areas where the Meadow Brown, and to a lesser extent the Gatekeeper, are able to thrive. In the more rural parts of the county though it is a common species, particularly in damper areas and in the shadow of woodland. Typical habitat includes a certain amount of scrub, and it is possible that it has increased since the advent of myxomatosis. The distribution map suggests that the chalk is the favoured geological formation, but this is probably due to recorder bias as recent detailed recording in the central weald has shown the butterfly to be present in most areas examined. Unlike most species of butterfly which are fastidious in their choice of oviposition site, the Ringlet female scatters her eggs at random when in flight. The larvae are frequently found when sweeping in the spring. Aberration occurs most frequently in the spotting and "ringlets" of the underside. Those forms with reduced or absent spots, ab. *parvipuncta* Castle Russell, ab. *arete* Müller, and ab. *caeca* Fuchs are not uncommon in certain localities, while ab. *lanceolata* Shipp, in which the spots are elongated, does not seem to have been recorded in the wild. Examples having the recessive gene for this form were released in a field near Cranleigh, where examples were seen for a few years afterwards, but not apparently recently.

Coenonympha pamphilus (Linnaeus, 1758) PLATE 14 **Small Heath**
ssp. *pamphilus* (Linnaeus, 1758)

Resident; grassland, woodland rides, heathland; fairly widespread and common.

Bi- or trivoltine; overlapping broods, late May to early September.

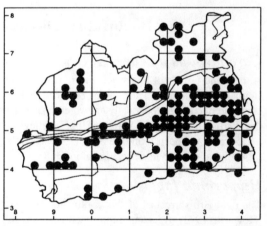

Nationally the Small Heath is one of our most ubiquitous butterflies, being found the length and breadth of the country in a variety of habitats and from sea level to the tops of mountains. When examined at a more local level however its status must be reassessed; in Surrey it is only ubiquitous on the chalk. On the commons of the London clay and the woodland rides and roadsides of the weald, it is rather less frequent, and on the sandy areas it is quite restricted. The adult may be seen throughout the flight period and there are usually several overlapping broods. Unlike other species of Brown, the larvae are hardly ever encountered when sweeping.

Danaus plexippus (Linnaeus, 1758) PLATE 16 **Monarch**
Migrant and escape from captivity.

The Monarch butterfly is a migrant species in this country, which certainly arrives from North America on occasion and may also originate from the Canary Islands where it has been resident for over a century and from where is currently in the process of establishing itself in southern Spain. In addition it is regularly bred in the many butterfly houses around the country and its most recent appearance in Surrey is due to a number of individuals escaping from such a source. The same year (1981) also saw the largest migration to this country ever recorded, probably in excess of 130 individuals. Their north American origin was supported by the arrival at the same time of several species of North American bird and by the technique of climatic backtracking. This westerly migration did not reach Surrey.

19th century – **Newlands Corner**, July 1896 (South, 1906).

1947 – **Dorking**, 29.5 by Buckstone (*Ent.* **81**:111).

1948 – 16.5 **Vachery Lake, Cranleigh**, seen by Rev Johnstone (Kettlewell, *Ent.* **81**:152).

1957 – **Norbury**, 25.9 (Waller, *Ent.* **90**:307).

1967 – **Godstone**, 21.10 and 29.10 (Plant, 1987).

1973 – **Oxted**, 16.10 (Chalmers-Hunt, *Proc. BENHS* **16**:3).

1978 – **Epsom**, 11.9 (Holdaway, *Ent. Rec.* **91**:27).

1981 – **Kew Gardens** and surrounds, 10-15.8, several adults seen and ova and larvae found on milkweed plants (*Asclepias incarnata, A. syriaca* and *A. speciosa*) – the result of an escape from a nearby butterfly house (Keesing, 1982).

Doubtful species and vagrants

Carcharodus alceae (Esper, 1780) — Mallow Skipper

A male and female of this species were taken in Surrey in June, 1923, by Baron Bouck. Being in good condition and having been caught flying around plants of mallow they were considered to have bred locally and when Frohawk described them, he bestowed on them the English name of The Surrey Skipper, believing the species to be a resident. That no more have ever been found, in Surrey or elsewhere in England, indicates that they were probably no more than an accidental introduction.

Hipparchia fagi (Scopoli, 1763) — Woodland Grayling

The only British record of this continental species is of one taken at Oxted, July 1946 (Howarth, 1973). Accidental introduction was considered the most likely explanation of its presence here. In Europe its distribution is central and southern, being absent from the north of France.

Arethusana arethusa ([Denis and Schiffermüller], 1775) — False Grayling

The only British record is of a single male found amongst a series of Grayling taken near Ash Vale, 21.8.1974 by A.J. Hedger (Hedger, *Ent. Gaz.* **28**:73). It may have been migrant as it is resident in north-western France, but is more likely to have been accidentally introduced.

Carterocephalus palaemon (Pallas, 1771) — Chequered skipper

Doubtful.

The Chequered Skipper is an extinct resident of the midland counties of England, and a resident in western Scotland. It is mentioned without comment in a list of the butterflies of the Reigate district (Brewer, 1856), which also includes such species as the Black-veined White, Black Hairstreak and Heath Fritillary, and also stated to have been taken at Mickleham in 1904 or 1905 (*Ent.* **39**:24). None of the subsequent authors mention it as a Surrey species and considerable doubt must attach to the accuracy of these records.

APPENDIX 1 – Plant list

agrimony	*Agrimonia eupatoria*	Rosaceae
alder buckthorn	*Frangula alnus*	Rhamnaceae
aspen	*Populus tremula*	Salicaceae
bilberry	*Vaccinium myrtillus*	Ericaceae
bird's-foot trefoil	*Lotus corniculatus*	Leguminosae
black currant	*Ribes nigrum*	Grossulariaceae
black medick	*Medicago lupulina*	Leguminosae
bracken	*Pteridium aquilinum*	Dennstaedtiaceae
bramble	*Rubus fruticosus* agg.	Rosaceae
broom	*Cytisus scoparius*	Leguminosae
buckthorn	*Rhamnus catharticus*	Rhamnaceae
buddleia	*Buddleia davidii*	Buddleiaceae
cabbage	*Brassica* spp.	Cruciferae
carrot	*Daucus carota*	Umbelliferae
charlock	*Sinapis arvensis*	Cruciferae
cock's-foot grass	*Dactylis glomerata*	Gramineae
common dog-violet	*Viola riviniana*	Violaceae
common nettle	*Urtica dioica*	Urticaceae
common rock-rose	*Helianthemum nummularium*	Cistaceae
common sallow	*Salix cinerea*	Salicaceae
common sorrel	*Rumex acetosa*	Polygonaceae
cowslip	*Primula veris*	Primulaceae
creeping cinquefoil	*Potentilla reptans*	Rosaceae
creeping thistle	*Cirsium arvense*	Compositae
cross-leaved heath	*Erica tetralix*	Ericaceae
cuckoo flower	*Cardamine pratensis*	Cruciferae
currant	*Ribes* spp.	Grossulariaceae
cut-leaved crane's-bill	*Geranium dissectum*	Geraniaceae
devil's-bit scabious	*Succisa pratensis*	Dipsacaceae
dogwood	*Cornus sanguinea*	Cornaceae
dyer's greenweed	*Genista tinctoria*	Leguminosae
elm	*Ulmus* spp.	Ulmaceae
English elm	*Ulmus procera*	Ulmaceae
everlasting pea	*Lathyrus latifolius*	Leguminosae
field pepperwort	*Lepidium campestre*	Cruciferae
firethorn	*Pyracantha coccinea*	Rosaceae
garlic mustard	*Alliaria petiolata*	Cruciferae
goat willow	*Salix caprea*	Salicaceae
gooseberry	*Ribes uva-crispa*	Grossulariaceae
gorse	*Ulex europaeus*	Leguminosae
hairy violet	*Viola hirta*	Violaceae

heather	*Calluna vulgaris*	Ericaceae
hedge mustard	*Sisymbrium officinale*	Cruciferae
hemlock water-dropwort	*Oenanthe crocata*	Umbelliferae
holly	*Ilex aquifolium*	Aquifoliaceae
honesty	*Lunaria annua*	Cruciferae
honeysuckle	*Lonicera periclymenum*	Caprifoliaceae
hop	*Humulus lupulus*	Cannabaceae
horse-radish	*Armoracia rusticana*	Cruciferae
horse-shoe vetch	*Hippocrepis comosa*	Leguminosae
ivy	*Hedera helix*	Araliaceae
kidney vetch	*Anthyllis vulneraria*	Leguminosae
mallow	*Malva* spp.	Malvaceae
marsh thistle	*Cirsium palustre*	Compositae
marsh violet	*Viola palustris*	Violaceae
meadow grass	*Poa* spp.	Gramineae
Michaelmas daisy	*Aster* spp.	Compositae
milkweed	*Asclepias* spp.	Asclepiadaceae
nasturtium	*Tropaeolum majus*	Tropaeolaceae
oak	*Quercus* spp.	Fagaceae
poplar	*Populus* spp.	Salicaceae
primrose	*Primula vulgaris*	Primulaceae
privet	*Ligustrum* spp.	Oleaceae
radish	*Raphanus* spp.	Cruciferae
red currant	*Ribes rubrum* agg.	Grossulariaceae
salad burnet	*Sanguisorba minor*	Rosaceae
sallow	*Salix* spp.	Salicaceae
sheep's-fescue	*Festuca ovina*	Gramineae
shepherd's purse	*Capsella bursa-pastoris*	Cruciferae
sloe	*Prunus spinosa*	Rosaceae
spear thistle	*Cirsium vulgare*	Compositae
tall melilot	*Melilotus altissima*	Leguminosae
timothy grass	*Phleum pratense*	Gramineae
tor grass	*Brachypodium pinnatum*	Gramineae
tormentil	*Potentilla erecta*	Rosaceae
Turkey oak	*Quercus cerris*	Fagaceae
violet	*Viola* spp.	Violaceae
viper's bugloss	*Echium vulgare*	Boraginaceae
wild strawberry	*Fragaria vesca*	Rosaceae
wood false-brome	*Brachypodium sylvaticum*	Gramineae
wych elm	*Ulmus glabra*	Ulmaceae

APPENDIX 2 – Gazetteer of sites

The gazetteer should enable sites mentioned in the text to be located within the county of Surrey. The map references are an approximation and should not be taken to imply that any particular record should be allocated to a particular 1 km square.

Addington	TQ3762	
Addiscombe	TQ3466	
Alfold	TQ0233	
Alice Holt	SU8241	Alice Holt itself is in N. Hants, but abuts the Surrey boundary
Ash Vale	SU8953	
Ashtead Common	TQ1759	Corporation of London
Bagmoor Common	SU9242	Surrey Wildlife Trust reserve
Bagshot	SU9163	
Balham	TQ2873	
Banstead Downs	TQ2561	
Banstead Heath	TQ2354	
Barnsthorns Wood	TQ0956	
Battersea	TQ2876	
Beddington	TQ2866	
Betchworth	TQ2051	
Bisley	SU9559	
Blackheath	TQ0345	
Bletchingley	TQ3250	
Bookham Common	TQ1256	National Trust
Botany Bay	SU9734	Forestry Commission
Box Hill	TQ1751	National Trust
Bramley	TQ0143	
Brentmoor Heath	SU9361	Surrey Wildlife Trust reserve
Brockham Warren	TQ1951	
Brook	SU9237	
Buckland	TQ2250	
Buckland Hills	TQ2352	
Bury Hill	TQ1548	
Camberley	SU8860	
Canfold Woods	TQ0739	
Carshalton	TQ2764	
Caterham	TQ3355	

Charterhouse	SU9645	
Cheam	TQ2463	
Chessington	TQ1863	
Chiddingfold	SU9635	
Chipstead	TQ2757	
Chobham	SU9761	
Chobham Common	SU9764	National Nature Reserve
Chobham, Longcross	SU9766	
Clandon Downs	TQ0550	
Clasford Common	SU9452	
Claygate	TQ1663	
Cleygate Common	SU9153	
Cobham	TQ1160	
Colekitchen Down	TQ0848	Surrey Wildlife Trust reserve
Colley Hill	TQ2452	National Trust
Combe Bottom	TQ0648	
Coombe Wood	TQ2070	
Cranleigh	TQ0638	
Croham Hurst	TQ3363	
Croydon	TQ3265	
Dawcombe	TQ2152	Surrey Wildlife Trust reserve
Devils Jumps	SU8639	
Dollypers Hill	TQ3158	Surrey Wildlife Trust reserve
Dorking	TQ1649	
Dulwich	TQ3373	
Dunsfold	SU9934	
Durfold Wood	SU9832	
Edolphs Copse	TQ2342	Woodland Trust
Effingham	TQ1153	
Egham	TQ0071	
Elm Corner	TQ0757	
Epsom	TQ2060	
Epsom Common	TQ1860	
Epsom Downs	TQ2158	
Esher Common	TQ1362	
Ewell	TQ2162	
Ewhurst	TQ0939	

Fairmile Common	TQ1161	
Fairoaks	TQ0062	
Farnham	SU8347	
Farthing Downs	TQ3057	Corporation of London
Fetcham Downs	TQ1554	
Field Common	TQ1366	
Fisherlane Wood	SU9832	Forestry Commission
Frensham Common	SU8440	National Trust
Friday Street	TQ1245	
Frith Hill	SU9058	
Gibbet Hill	SU8935	
Givons Grove	TQ1754	
Glovers Wood	TQ2240	Woodland Trust
Godalming	SU9743	
Gomshall	TQ0848	
Guildford	SU9949	
Hackhurst Downs	TQ0948	National Trust
Hale Common	SU8349	
Ham Common	TQ1871	
Hambledon	SU9638	
Hankley Golf Course	SU8842	
Happy Valley	TQ3057	
Haslemere	SU9032	
Headley Down	TQ1953	National Trust
Headley Heath	TQ1953	National Trust
Headley Warren	TQ1954	private land
Henleypark Lake	SU9353	
Hinchley Wood	TQ1565	
Hindhead	SU8937	National Trust
Hog Wood	TQ0132	
Hog's Back	SU9248	
Holmwood Common	TQ1745	National Trust
Horsell	TQ0060	
Horsley	TQ0952	
Howell Hill	TQ2362	Surrey Wildlife Trust reserve

Juniper Bottom	TQ1752	National Trust
Juniper Hill	TQ2352	National Trust
Kenley Common	TQ3358	Corporation of London
Kew Gardens	TQ1876	
Kingston	TQ1869	
Leatherhead	TQ1656	
Leigh	TQ2345	
Leith Hill	TQ1343	
Lightwater	SU9161	
Little Frensham Pond	SU8541	National Trust
Littlefield Common	SU9552	
Lower Canfold Wood	TQ0739	
Lucas Green	SU9459	
Merrow Downs	TQ0249	
Mickleham	TQ1753	
Milford	SU9441	
Mitcham Common	TQ2868	
Molesey	TQ1268	
Netley Heath	TQ0849	
Newlands Corner	TQ0449	
Norbury	TQ3169	
Norbury Park	TQ1653	
Oaken Wood	SU9933	Forestry Commission
Ockley Common	SU9141	
Onslow Village	SU9749	
Ottershaw	TQ0263	
Oxshott Heath	TQ1361	
Oxted	TQ3952	
Park Downs	TQ2658	
Pewley Downs	TQ0048	
Pirbright Common	SU9254	
Pitch Hill	TQ0842	
Pockford Bridge	SU9835	
Princes Coverts	TQ1561	private land
Puttenham Common	SU9045	

Ranmore (Common)	TQ1450	
Raynes Park	TQ2268	
Redhill	TQ2850	
Reigate	TQ2550	
Reigate Hill	TQ2552	National Trust
Richmond	TQ1874	
Richmond Park	TQ2072	
Riddlesdown	TQ3260	Corporation of London
Riddlesdown Quarry	TQ3359	
Ripley	TQ0556	
Royal Common	SU9242	
Selsdon	TQ3562	
Shackleford	SU9245	
Sheepleas	TQ0851	
Sheerwater	TQ0260	
Sidney Wood	TQ0234	Forestry Commission
South Croydon	TQ3363	
South Godstone	TQ3648	
South Hawke	TQ3753	National Trust
St. Ann's Hill	TQ0267	
St. Martha's Hill	TQ0248	
Staffhurst Wood	TQ4148	
Streatham	TQ3071	
Stringers Common	SU9952	
Surbiton	TQ1867	
Surrey Docks	TQ3679	
Sutton	TQ2564	
Tatsfield	TQ4157	
Tattenham Corner	TQ2258	
Thames Ditton	TQ1567	
Thorpe	TQ0067	
Thursley Common	SU9041	National Nature Reserve
Tilford	SU8743	
Tugley Wood	SU9833	Forestry Commission
Tulse Hill	TQ3173	
Tunnel Hill	SU9155	

Vann Lake	TQ1539	Surrey Wildlife Trust reserve
Virginia Water	SU9768	
Wallis Wood	TQ1238	Surrey Wildlife Trust reserve
Walton-on-Thames	TQ1066	
Warlingham	TQ3558	
Wentworth	SU9867	
Westcott Downs	TQ1349	National Trust; often incorrectly known as Ranmore Common
West Croydon	TQ3266	
West End, Esher	TQ1263	
Westend Common	SU9361	
Weybridge	TQ0764	
White Downs	TQ1148	National Trust
White Hill	SU9160	
Wimbledon Common	TQ2271	
Windlesham	SU9263	
Winkworth Arboretum	SU9941	National Trust
Wisley Common	TQ0658	
Wisley RHS	TQ0658	
Witley	SU9439	
Witley Common	SU9240	National Trust
Woking	TQ0058	
Woldingham	TQ3656	
Worcester Park	TQ2265	
Wormley	SU9438	
Worplesdon Hill	SU9653	
Wotton	TQ1346	
Wyke Common	SU9152	

APPENDIX 3 – Organisations

Study societies

British Entomological and Natural History Society
Dinton Pastures Country Park, Davis Street, Hurst,
Reading, Berks RG10 0TH.

Croydon Natural History and Scientific Society
96a Brighton Road, South Croydon, Surrey CR2 6AD.

Conservation bodies

Butterfly Conservation
Surrey branch – S. Jeffcoate, P.O. Box 188, Dorking, Surrey RH4 1YT.
Head office – P.O. Box 222, Dedham, Colchester, Essex, CO7 6EY.

Surrey Wildlife Trust
School Lane, Pirbright, Surrey GU24 0JN.

APPENDIX 4 – References

Allan, P.B.M., 1980.
Leaves from a moth-hunter's notebook. Classey, Faringdon.

Asher, J., 1994.
The Butterflies of Berkshire, Buckinghamshire and Oxfordshire. Pisces.

Baker, P.J., 1986.
Changes in the status of the lepidoptera of a north west Surrey locality. *Proc. Trans. Br. ent. nat. Hist. Soc.* **19**:33.

Barrett, C.G., 1893.
The Lepidoptera of the British Islands, **1**. Reeve, London.

Bretherton, R.F., 1951a.
The early history of the Swallow-tail butterfly (*Papilio machaon* L.) in England. *Entomologist's Rec. J. Var.* **63**:206.

Bretherton, R.F., 1951b.
Our lost Butterflies and Moths. *Entomologist's Gaz.* **2**:211.

Bretherton, R.F., 1957.
A List of the Macrolepidoptera and Pyralidina of north-west Surrey. *Proc. S. Lond. ent. nat. Hist. Soc.* **1955**:94-151.

Bretherton, R.F., 1965.
Additions to the List of Macrolepidoptera and Pyralidina of north-west Surrey.
Proc. S. Lond. ent. nat. Hist. Soc. **1965**:18- 30.

Brewer, J.A., 1856.
A new flora of the neighbourhood of Reigate, Surrey. William Pamplin, London.

Buckell and Prout, 1898-1901.
The fauna of the London district: Lepidoptera. *Trans. City of London Entomological and Natural History Society.* **8**:51-63; **9**:66-80; **10**:62-74.

Chalmers-Hunt, J.M., 1960-61.
The Butterflies and Moths of Kent, **1**. Arbroath and London.

Chevallier, L.H.S., 1952.
Lampides boeticus Linn. in Surrey.
Entomologist's Rec. J. Var. **64**:274.

Collier, A.E., 1959.
A forgotten discard: the problem of redundancy.
Entomologist's Rec. J. Var. **71**:118.

Dandy, J.E., 1969.
Watsonian Vice-counties of Great Britain.
Ray Society, London.

de Worms, C.G.M., 1950.
The Butterflies of London and its Surroundings.
Lond. Nat. **29**:46-80.

de Worms, C.G.M., 1959.
A Supplement to the Butterflies and Moths of London and its Surroundings; part 1. *Lond. Nat.* **38**:33-39.

Emmet, A.M. (ed.), 1989.
The Moths and Butterflies of Great Britain and Ireland, **7(1).**
Harley Books, Essex.

Evans, L.K., and Evans, K.G.W., 1973.
A Survey of the Macrolepidoptera of Croydon and north-east Surrey. *Proc. Croydon Nat. Hist. Sci. Soc.* **XIV**:273-408.

Goater, B., 1974.
The Butterflies and Moths of Hampshire and the Isle of Wight. Classey, Faringdon.

Goater, B., 1992.
The Butterflies and Moths of Hampshire and the Isle of Wight: additions and corrections. Joint Nature Conservation Committee.

Goss, H., 1902.
Butterflies and Moths. *The Victoria History of the County of Surrey,* **3** *Zoology.* Constable, London. [*VCH*]

Howarth, T.G., 1973.
South's British Butterflies. Warne, London.

Keesing, J.L.S., 1982.
Monarch butterflies – *Danaus plexipus (sic)* – at Kew.
Bull. amat. Ent. Soc. **41**:74.

Morris, R.K.A., and Collins, G.A., 1991.
On the hibernation of Tissue moths *Triphosia dubitata* L. and the Herald moth *Scoliopteryx libarix* L. in an old fort.
Entomologist's Rec. J. Var. **103**:313.

Morton, A.J., and Collins, G.A., 1992.
Distribution analysis of Surrey Lepidoptera using the DMAP computer package. *Nota lepid.* **15** (1):84-88.

Newman, E., 1874.
The Illustrated Natural History of British Butterflies.
Hardwicke, London.

Oldacre,F.A., 1913.
A list of the Lepidoptera occuring within six miles of Haslemere. Haslemere Natural History Society.

Oates, M.R., and Warren, M.S., 1990.
A review of butterfly introductions in Britain and Ireland.
Joint Committee for the Conservation of British Insects.

Perrins, C.M., 1959. in Anon.
Birds Butterflies Moths of the Godalming district.
Charterhouse Natural History Society.

Philp, E.G., 1993.
The Butterflies of Kent. Kent Field Club, Sittingbourne.

Plant, C.W., 1987.
The Butterflies of the London Area. London Natural History Society.

Pratt, C., 1983.
A modern review of the demise of *Aporia crataegi* L.: the Black-veined White.
Entomologist's Rec. J. Var. **95**:45, 161, 232.

Shirt, D.B., ed., 1987.
British Red Data Books: 2. Insects. Nature Conservancy Council, Peterborough.

South, R., 1906.
The Butterflies of the British Isles. Warne, London.

Stephens, J.F., 1828.
Illustrations of British Entomology (Haustellata), **1**. London.

Thomas, J., and Lewington, R., 1991.
The Butterflies of Britain and Ireland. Dorling Kindersley.

Tutt, J.W., 1905.
A Natural History of the British Lepidoptera, **8**.
Swan Sonnenschein, London.

Tutt, J.W., 1907.
A Natural History of the British Lepidoptera, **9**.
Swan Sonnenschein, London.

Tutt, J.W., 1908.
A Natural History of the British Lepidoptera, **10.**
Swan Sonnenschein, London.

Wheeler, A.S., 1955.
A Preliminary List of the Macro Lepidoptera of Bookham
Common. *Lond. Nat.* **34**:28.

APPENDIX 5 – Glossary

aposematic Carrying coloration, usually reds and yellows, that warns potential predators of danger. In a more general usage, it is also applied to warning scents, sounds and behaviour.

costa The leading edge of a wing.

cremaster A specialized structure of hooks and spines on the tail end of the pupa.

cryptic Hidden, camouflaged by shape or coloration.

gynandromorph An adult insect exhibiting both male and female characteristics; in the case of butterflies this is usually expressed as differences in wing colour or pattern. A bilateral gynandromorph is divided down the mid-line, one side being of male structure, the other, female.

beating A technique for finding larvae which feed on trees and shrubs, which involves tapping the branches with a stick and catching any dislodged larvae on a sheet or specially designed tray.

RDB status Red Data Book. Species recorded from fifteen or fewer 10 km squares are accorded RDB status. They are further categorized as RDB1 (Endangered), RDB2 (Vulnerable) or RDB3 (Rare) according to the perceived degree of threat to them rather than their actual rarity. See Shirt, 1987.

sweeping A technique for finding larvae which feed on low plants such as grasses and heather, which involves a heavy duty net bag which is swept rapidly through the vegetation.

INDEX

Figures in bold indicate plate numbers

INDEX

Figures in bold indicate plate numbers